The
Skulking Way
of War

The
Skulking Way
of War

TECHNOLOGY AND TACTICS AMONG THE NEW ENGLAND INDIANS

Patrick M. Malone

JOHNS HOPKINS UNIVERSITY PRESS

Baltimore and London

Originally published in 1991 in a hardcover edition by Madison Books,
Lanham, Maryland. Published in cooperation with Plimoth Plantation,
Plymouth, Massachusetts

Johns Hopkins Paperbacks edition, 1993

The Johns Hopkins University Press
2715 North Charles Street
Baltimore, Maryland 21218-4319
The Johns Hopkins Press Ltd., London

Library of Congress Cataloging-in-Publication Data

Malone, Patrick M.
    The skulking way of war: technology and tactics among the New England
Indians / Patrick M. Malone. — Johns Hopkins paperbacks ed.
        p.   cm.
    Originally published: Lanham, Md. : Madison Books, c1991.
    Includes bibliographical references and index.
    ISBN 0-8018-4554-8 (pbk. : acid-free paper)
    1. Indians of North America—New England—Wars. 2. Indians of North
America—New England—Material culture. 3. Warfare, Primitive—New
England—History. 4. Guerrilla warfare—History. 5. Technological inno-
vations—New England—History.   I. Title
[E78.N5M35   1993]
399' .08997074—dc20   92-22740

A catalog record of this book is available from the British Library.

FOR LYN
Cartographer
Steady directional compass

Contents

The
Skulking Way
of War

A Few Words on Skulking

After the Bloody Brook Massacre in King Philip's War, William Hubbard criticized the English for not fighting in a tight body as musketeers were supposed to do. He warned against any imitation of Indian tactics: ". . . skulking behind trees and taking . . . aim at single persons."

At the end of King Philip's War in 1677, John Eliot wrote: "In our first war with the Indians [The Pequot War of 1637], God pleased to show us the vanity of our military skill, in managing our arms, after the European mode. Now we are glad to learn the skulking way of war."

Neil Sheehan said that the American military advisors in Vietnam in 1962 shared one wish: "They hoped that the guerrillas would one day be foolish enough to abandon their skulking ways and fight fairly in a stand-up battle."*

* Quotation from Neil Sheehan, *A Bright Shining Lie: John Paul Vann and America in Vietnam* (New York: Random House, 1988), p. 204.

Introduction

This study is primarily a history of technological change, but it is also an examination of tactics and of the concept of total war. Technology has played an important role in warfare since men first tried to harm each other with rocks or clubs. It has always been a part of any military system and has influenced the conduct of all wars. The natural environment in which men fight has also had a strong effect on their military systems, determining in many cases the choice or design of military equipment and the ways in which it was used. Warfare has often been limited in its scope or ferocity by the deliberate restraint of combatants and by the capabilities of their technology. In southern New England in the seventeenth century, Europeans introduced unrestrained, total warfare and a well-developed technology for conducting it. Indians adopted much of this technology and then modified their traditional patterns of forest combat in response to the European example of total war.

During the period of this study, 1600–1677, Indians and white men fought two major wars in New England. The earliest conflict, a series of skirmishes and punitive expeditions called the Pequot War, took place on Block Island and the Connecticut coast in 1636 and 1637. The next war was on an entirely different scale; King Philip's War engulfed almost all of southern New England between 1675 and 1676. There is a good summary of those wars and their historical significance in Wilcomb Washburn's article on "Seventeenth-Century Indian Wars," which can be found in Bruce Trigger, ed., *Northeast*, volume 15 of the well-known

Handbook of North American Indians, edited by William Sturtevant and published in 1978 by the Smithsonian Institution, Washington, D.C.

The Pequot War pitted English colonists and their Indian allies against a single tribe that was accused, probably falsely, of murdering two white traders. The Pequots had alienated neighboring Indians and stood in the way of English expansion in Connecticut. The ease with which the English crushed the Pequots and the ruthless nature of their reaction to this "Indian problem" affected relations between the races for decades. Indians understood the potential of firearms in warfare after seeing the results of this unequal contest. For more information on the causes and implications of the Pequot War, see Neal Salisbury, *Manitou and Providence: Indians, Europeans, and the Making of New England, 1550–1648* (New York: Oxford U. Press, 1982).

By 1675, much of the native land of the Indians was in English hands. Colonial authorities subjected even tribal sachems to the rigorous demands of English law, and a way of life in the forests was slipping away. The humiliation of English domination and the threat of cultural extinction drove the great Wampanoag sachem Metacomet (called King Philip by the English) to an act of desperate resistance. The insurrection which began in June of 1675 spread to other tribes and has become known as King Philip's War. Historians now doubt that a widespread intertribal conspiracy was responsible for the conflict. They blame the heavy-handed actions of colonial forces for driving some neutral or wavering Indian groups into open hostility as the war expanded. The English eventually defeated the insurgent tribes, with the critical assistance of Indian allies, but the terrible struggle produced heavy casualties on both sides and left over a dozen towns in ashes. Indians with firearms had proven to be fearsome adversaries in a new type of forest warfare that drew inspiration from both cultures.

This book discusses only warfare in southern New England. King Philip's War also affected both Maine and New Hampshire. For detailed coverage of the war, see the standard narrative history by Douglas Leach, *Flintlock and Tomahawk: New England in King Philip's War* (New York: Norton, 1966) and a newly-published study by Russell Bourne, *The Red King's Rebellion: Racial Politics in New England, 1675–1678* (New York: Atheneum, 1990).

There is a great deal of source material on the English military system in New England and many English accounts of warfare between the colonists and the Indians. Learning the Indian side of the story is more difficult. Any detailed description of the southern New England Indians will always be incomplete. They left no written record of their culture, and those early explorers and colonists who tried to describe the way of life of these Native Americans were usually poor ethnographic observers. Archaeological evidence has increased our knowledge of these Indians and will continue to do so, but much of their history and culture will always remain a mystery. In studying the tribes of southern New England, a scholar must sometimes draw inferences from known practices of neighboring tribes and from accepted anthropological conclusions about the coastal Algonquian culture in general. He or she must also use observations made of the New England Indians after their culture had been significantly changed by contact with Europeans and European trade goods. Many cultural traits remained stable despite European contact and, with proper caution, can be considered evidence of an aboriginal practice.

Many institutions and individuals have helped the author in the preparation of this study. My interest in military history was stimulated by seminars at the U.S. Naval Academy with Professor W.H. Russell. Although I do not recommend participant observation as a research method for historians who want to study combat, the lessons that I learned during my service as a Marine in Vietnam have expanded my understanding of warfare in the forest.

My interest in this particular subject began with a seminar paper for Professor Carl Bridenbaugh at Brown University. I later wrote my doctoral dissertation on "Indian and English Military Systems in New England in the Seventeenth Century," working under the able direction of Professor A. Hunter Dupree at Brown University. Murray Murphey, editor of the *American Quarterly*, accepted my 1973 article on "Changing Military Technology Among the Indians of Southern New England, 1600–1677." I appreciate that journal's permission to use sections of the article in this publication.

Two periods of concentrated effort on the manuscript were made possible by the generosity of my employers; Brown University gave me a faculty summer fellowship, and Slater Mill Historic Site provided a month of sabbatical leave from my position as museum director. In writing this monograph, I have profited greatly from the advice and comments of Hunter Dupree, Gordon Wood, James Deetz, Neal Salisbury, and Richard Ehrlich. They are responsible for many of the best things in this study and for none of its flaws. Any errors of fact or interpretation are mine alone.

The superb illustrations done for this monograph by P.D. Malone, Lyn Malone, and David Macaulay have greatly enhanced the educational value and the attractiveness of the original publication. Nanepashemet, a Wampanoag Indian and a research associate at Plimoth Plantation, helped with the conception and design of David Macaulay's highly-detailed cover illustration for the original volume.

The Haffenreffer Museum of Anthropology at Brown University, the Peabody Museum at Harvard University, the George Hail Free Library, and the Robbins Museum of Archaeology kindly allowed the drawing of artifacts in their collections. I also want to thank the John Carter Brown Library, the John Hay Library, the Anne S.K. Brown Military Collection, the Rhode Island Historical Preservation Commission, the Essex Institute, Plimoth Plantation, the Museum of the American Indian, the George Hail Free Library, the Massachusetts Archaeological Society, the Museum of Art at the Rhode Island School of Design, the Haffenreffer Museum, Plimoth Plantation, and the New York Public Library for permission to reproduce illustrations.

The staffs of the following libraries, museums, and historical agencies provided access to the sources, both documentary and artifactual, that were used in the research for this study: the Brown University Libraries, the Harvard University Libraries, the Van Pelt Library of the University of Pennsylvania, the Massachusetts State Archives, the Massachusetts Historical Society, the Connecticut State Library, the Rhode Island Historical Society, the George Hail Free Library, the Rhode Island Historical Preservation Commission, the Haffenreffer Museum of Anthropology, the Robbins Museum of Archaeology, Pilgrim Hall, the Old Colony Historical Society, Plimoth Plantation, Fruitlands Museums, the Higgins Armory Museum, Heritage Plantation, the Peabody Museum at Harvard University, Fort Ticonderoga, and Springfield Armory National Historical Site.

Harold L. Peterson showed me his personal collection of colonial weapons and shared his ideas on the use of firearms in New England. Robert Gordon offered his insights on metal-working processes, gun repairs, and the composition of metal artifacts. In my continuing work on Indian technology, I hope to do more of the metallurgical testing that Professor Gordon has proposed.

A number of people helped with the original edition published by Madison Books. For editorial help, I am indebted to James Lyons, Mark McDonough, Donna Curtin, Carol Frost, Greg Galer, and Cooper Abbott. Marie Pelletier and Die Modlin oversaw every step of the original book's design.

Ted Avery, Ted Curtin, Richard Hurley, Brooke Hammerle, Kathy Franz, and Bill Rice did excellent photographic work. Ann Jerome, Dorothy Garceau, Cooper Abbott, Linda Elbers, and Peter Harrington found many of the illustrations used in the book. George Malone, who has since been killed in a tragic kayak incident, helped with research on colonial laws. Alasdair MacPhail saved for me everything he found on Indian and English military systems as he did his own doctoral research on seventeenth-century New England.

The many fine scholars whose publications influenced my work are credited in the citations for each chapter. My debt to them is enormous. I have also benefitted from the informal encouragement, constructive criticism, and suggestions of fellow students of Indian-White relations; I owe much to Neal Salisbury, Pat Rubertone, Paul Robinson, Nanapashemet, William Simmons, James Axtell, Anthony Wallace, Richard Slotkin, Michael Raber, Kevin McBride, Myron Stachiw, Richard Colton, Jeremy Bangs, Larry Babits, Nancy Lurie, Wilcomb Washburn, Adam Hirsch, Gail Gustafson, Alden Vaughan, Ella Sekatau, Len Travers, James W. Baker, and Francis Jennings.

Numerous drafts, revisions, and changes in the original manuscript would not have been possible without the skillful typing and general assistance of Judith Malone Neville, Winifred Barton (now deceased), Sandra Norman, and Fay Martineau. Dr. Neville also helped with the citations. My wife, Lyn, has been a constant source of support and good cheer.

James Deetz first suggested that I prepare a manuscript on this topic for Plimoth Plantation. He deserves a great deal of credit for getting the project underway. At his urging the Society of Colonial Wars generously provided a grant to promote publication. I am particularly grateful to Richard Ehrlich, the prime mover of this project at Plimoth Plantation.

In 1991, Merritt Roe Smith recommended that Johns Hopkins University Press publish *The Skulking Way of War* in paperback. I thank him for his confidence in my work and editor Robert J. Brugger for helping to make this new edition a reality.

Patrick M. Malone

A Note on Spelling, Punctuation, and Dating

The author has modernized the spelling of words in quotations from the seventeenth-century sources but has changed the original punctuation only when necessary to clarify the meaning of a passage.

The dates used in this study also reflect a modest departure from seventeenth-century practice. In the Julian Calendar of that period, a new year did not begin until March 25. To avoid any confusion among modern readers, the author has chosen to put all dates from January 1 through March 24 into the new year rather than the previous one.

A Comment on Historic Illustrations

Many of the illustrations for this book are taken from popular histories published in the colonial period or the nineteenth century. The old drawings may not be accurate in every detail. Some reveal the biases of illustrators, publishers, or authors. Others give a loose interpretation of events and a fanciful re-creation of participants. Despite these weaknesses, there is much to be gained from examination of the historic illustrations. They are particularly useful in evoking the horrors of total warfare in early New England, and they often depict a tactical situation with considerable realism. Several of the drawings of craftsmen at work are valuable acknowledgments of Indian technological abilities. The caption for each figure and the Notes to Illustrations will provide additional information and guidance.

Abbreviations in Citations

CA	Connecticut State Archives.
CCR	J. Hammond Trumbull, ed., *Public Records of the Colony of Connecticut* (15 vols.; Hartford, Conn., 1850–1890).
MA	Massachusetts State Archives.
MCR	Nathaniel Shurtleff, ed., *Records of the Governor and Company of the Massachusetts Bay in New England* (5 vols.; Boston, 1853–1854)
MHC	*Collections of the Massachusetts Historical Society.*

NHCR   Charles J. Hoadly, ed., *Records of the Colony and Plantation of New Haven* (2 vols.; Hartford, Conn., 1857-1858).

PCR   Nathaniel Shurtleff and David Pulsifer, eds., *Records of the Colony of New Plymouth* (12 vols.; Boston 1855–1861).

RICR   John R. Bartlett, ed., *Records of the Colony of Rhode Island and Providence Plantations in New England* (10 vols.; Providence, R.I., 1856–1865).

RIHC   *Collections of the Rhode Island Historical Society.*

Fig. 1. Romantic view of the New England forest.

THE SKULKING WAY OF WAR

CHAPTER I

The Aboriginal Military System

At the beginning of the seventeenth century, the Indians living in the area between the Merrimack River and Long Island Sound shared many cultural traits. They were all inhabitants of the woodlands, and their several dialects were part of the Algonquian language family. Similar weapons, fortifications, tactics, and attitudes toward warfare existed in each tribe of this region, which we now call southern New England.[1] The traditional technology of the Indians not only protected them from the challenges of a harsh natural environment, but also prepared them to meet their human enemies. Frequent conflict and fleeting alliances linked the tribes in a dynamic system of organized violence, a system dependent on technology. The ancient arrow maker and the woman who prepared rations for a war party were as essential in the operation of the military system as the warriors who did the fighting.

Feuds between kin groups and intertribal or interband wars of varying scale and intensity were common. Combat was usually on a small scale, however, with ambushes and raids on villages much more frequent than actual battles. Indians did undertake prolonged sieges of fortified positions on occasion and would sometimes meet in open fields for battles or skirmishes involving large numbers of warriors. In all these forms of warfare, relatively few participants were ever killed.[2]

Reasons for warfare were many and were sometimes so complex and intertwined that the participants themselves may not have understood them. New England Indians fought to gain prestige and power, to demonstrate their courage and martial skills, to resist aggression, to dominate weaker neighbors, to extort tribute, to gain hunting territory and fishing rights, to control trade, and to avenge real or imagined wrongs. Daniel Gookin, a Puritan missionary, believed that vengeance was a primary motive for combat between Indians. However, the violence which warriors directed against "such as have injured them or their kindred" was sometimes an effective social mechanism for the easing of severe grief and cultural stress within a band. Damage to group or individual pride also demanded some type of response and could lead to open warfare. Roger Williams reported that "mockery (between their great ones)" was "a great kindling of war amongst them." Long-standing grievances which European observers blamed for much of the armed conflict among Indian groups could provide ready justification for military action prompted by newer issues, such as land depletion and the increasing competition for access to European trade goods in the mid-seventeenth century.[3]

The leader of a band, known as the sachem, encouraged his or her followers to maintain a state of readiness for warfare. The principal sachem of a tribe relied on the advice of a council and needed the support of other sachems who controlled bands within the tribe. His authority over these bands was not secure, and they might break away to establish their independence. Often the name which the English used to distinguish a tribe was actually the name of just one band within a loose confederation of many bands. Several separate tribes could also form alliances, with one of the sachems exercising a tenuous authority over all. The power of the sachems varied greatly in New England, and English observers had great difficulty in comprehending the limits of their authority. It is clear that sachems had no hereditary right to their positions, that they had to cultivate the support of their followers, and that they could be deposed.[4] In time of war, leadership frequently shifted to particular warriors who had demonstrated their ability to conduct successful military operations.[5]

The sizes of the major tribes and the boundaries of the areas they controlled before the introduction of devastating European diseases are the subjects of much historical debate. Recent scholarship has suggested that previously accepted estimates of Indian population figures must be revised upward. It seems likely that over seventy-five thousand Indians lived in what is now New England at the beginning of the seventeenth century. Most of these aboriginal inhabitants occupied the coastal plains and river valleys in the southern half of the region.[6]

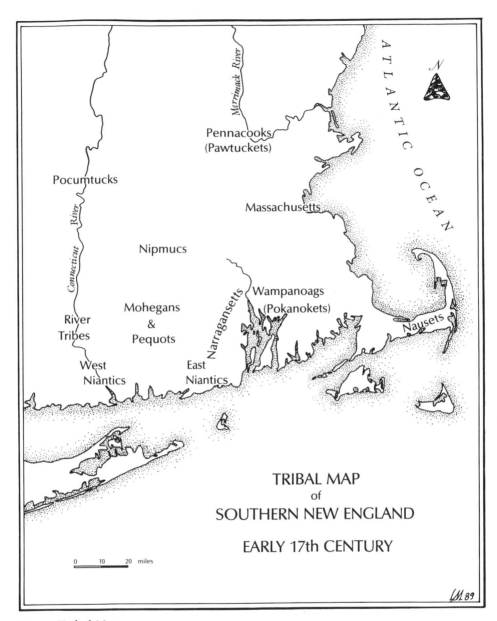

ATLANTIC OCEAN

Merrimack River

Pennacooks
(Pawtuckets)

Pocumtucks

Massachusetts

Connecticut River

Nipmucs

Wampanoags
(Pokanokets)

River
Tribes

Mohegans
&
Pequots

Narragansetts

Nausets

West
Niantics

East
Niantics

TRIBAL MAP
of
SOUTHERN NEW ENGLAND

EARLY 17th CENTURY

0   10   20   miles

LM 89

Fig. 2. Tribal Map.

The three largest tribes in southern New England were the Massachusetts, the Wampanoags (also called Pokanokets), and the Narragansetts. There may have been twelve thousand Massachusetts living near the bay of the same name. South of them were the various bands of the Wampanoags, a tribe of about the same size as the Massachusetts. Their lands stretched from the base of Cape Cod to the eastern shore of Narragansett Bay, and their influence extended over the neighboring Nausets of Cape Cod and the island-dwelling Indians on Nantucket and Martha's Vineyard. The Narragansetts, frequently in conflict with the Wampanoags, had the greatest population of all; there were perhaps sixteen thousand Narragansetts in most of what is today the state of Rhode Island.[7]

Many tribes lived in the area which we know as Connecticut. The cluster of bands near the Connecticut River are usually called the River Tribes. Two major divisions of the Niantics held much of the southern coast, with the territory of the Eastern Niantics extending to the Pawcatuck River in Rhode Island. The Pequots and their allied tribe the Mohegans claimed a large region between and north of the Eastern and Western Niantics.[8]

The Pocumtucks were named for the largest of a group of bands on the Connecticut River in what is now western Massachusetts. Another tribe, the Nipmucs, controlled most of the woodlands between the Pocumtucks and the Massachusetts. North of these tribes were many villages of the Pennacooks (also called Pawtuckets) who lived along or near the Merrimack River. To the west of the New England tribes, along the Hudson River and beyond, were the powerful Mohawks, the closest of the five Nations of the Iroquois. In the seventeenth century, the Mohawks became deeply involved in the political, military, and economic affairs of southern New England tribes.[9]

All of the above tribes were primarily agricultural, but farther north the conditions for agriculture were not as favorable. The Abenaki bands, groups of northern Algonquians who hunted and fished throughout part of present New Hampshire and Maine, were much less involved with the cultivation of crops. Most of these Indians belonged to a distinct culture area where the demands of hunting produced a way of life different from that in southern New England. The Abenaki sometimes fought with the Massachusetts in the early seventeenth century but also traded for the agricultural surpluses of the southern tribes.[10]

Agriculture greatly affected the military patterns in southern New England. Maize, or corn, could be dried and stored in the ground before an expected war, thereby providing a relatively secure source of provisions for a year or more of fighting. The annual harvests of beans, squash, and corn also gave stability and cohesion to villages and bands. The agricultural Indians did not have to spend most of the year struggling for survival in small, isolated hunting groups. Men hunted frequently to supplement the Indian diet, but women in the fields supplied most of the food and gave men the leisure time necessary to organize and carry out extended warfare.[11]

The maize which kept Indian families from starving during long wars was equally valuable as a lightweight combat ration for warriors traveling great distances. Each man on a military expedition carried a bag of powdered, dry corn called "nocake." Roger Williams said that "with this ready provision, and their bows and arrows, are they ready for war, and travel at an hour's warning." The durable and nourishing powder eliminated the need for cooking fires which might alert an enemy. Warriors nearing the territory of their opponents simply mixed nocake with water and ate the corn paste without risky delays.[12]

Mobility was critical in Indian warfare, but rapid travel in the New England environment required some technological aids. Moccasins, manufactured from the supple skins of deer or the tougher hides of moose, were lightweight, comfortable, and quiet in the forest. When winter snows blanketed the region, an Indian strapped snowshoes to his moccasined feet and moved freely through the deepest drifts after game or enemies.[13]

Bodies of water were no serious challenge to movement. Indians used two types of canoes in southern New England. The birchbark canoe was known even in regions where native birches did not provide suitable bark. Materials or entire vessels were traded between tribes, making this superbly-designed, light canoe form available over a broad area. A much heavier canoe, carved from the trunk of a tree, was more common among the Wampanoags, Narragansetts, and various Connecticut tribes. Such "dug out" canoes of pine, oak, or chestnut were not easy to carry overland on portages between navigable waters, but they were stable enough to handle even the rough waves and swells of coastal bays and sounds. Some were up to fifty feet long and could carry twenty Indians, according to Daniel Gookin. Williams reported that he had "known thirty or forty of their canoes [to be] filled with men, and [known of] near as many more [canoes] of their enemies in a seafight."[14]

Warfare often required movement of the entire population of a village or camp with very little notice. Sachems, warned of a possible attack, could gather separate groups into one or more easily defensible positions or temporarily scatter their people into mobile bands capable of avoiding their enemies. Southern New England Indians were not nomads, but they were used to seasonal moves and could pack their belongings quickly. With a few hours' notice, an entire village could be broken up and the inhabitants on their way to one or more places of refuge.[15]

Indians lived in several different places during an average year. They could easily disassemble and transport their houses, which they made of sewn cattail mats or bark panels on a framework of poles. Large groups dwelled in villages near cleared fields during most of the warm months and spent the winters in sheltered, wooded valleys. Indians also congregated at good fishing spots in the spring, but went off in small groups to fall hunting camps. At some of their seasonal village sites they built forts for protection against their enemies.[16]

William Wood explained that "they made themselves forts to fly into, if the enemies should unexpectedly assail them." Their defenses were palisades of upright logs, "ten or twelve foot high, rammed into the ground, with undermining within, the earth being cast up for their shelter against the dischargements of their enemies." Entrances, made by overlapping ends of the palisade walls, could be closed with brush or other obstructing material. In most cases these forts were round, although square and rectangular forms were also used. A group might enclose one or more houses or an entire village within a single fort. In time of danger, the Indians living near such a place of refuge could move closer to the palisade or erect temporary shelters within it. Imminent warfare spurred increased construction of these vital defensive fortifications.[17]

The fort was an effective shield in the limited scope of New England aboriginal warfare, where such a weapon as fire was considered too horrible and deadly to use. From within the walls of their forts, defending warriors relied on their bows and arrows to keep an assaulting force at bay. Wood said that the archers used the gaps between logs in the palisades "to send out their winged messengers, which often delivered their sharp and bloody embassies

Fig. 3. Birchbark canoe.

Fig. 4. Making a dug out canoe. An engraving of a 1585 painting by
Roanoke colonist John White.

Fig. 5. The fortified, coastal Algonquian town of Pomeiooc. An engraving of a 1585 painting by Roanoke colonist John White.

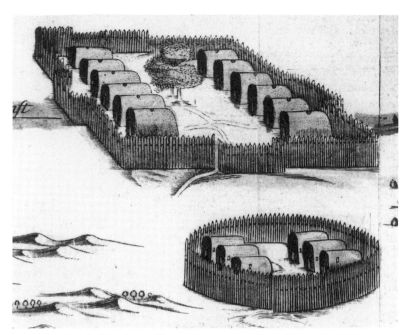

Fig. 6. Palisaded Indian villages shown on several maps by John Seller, including his map of New England in 1676.

in the tawny side of their naked assailants, who, wanting butting rams and battering ordinances to command at distance, lose their lives by their too near approachments."[18]

Few Indians were willing to charge into a hail of arrows and climb over a well-defended palisade. As a result of their realistic assessment of the personal danger involved in an all-out assault, the attacking warriors often chose the safer alternative of a long-term siege. They shot at the men behind the palisade and hurled insults in an effort to make the defenders come out and fight. Occasionally the taunts of the attackers did bring on an open battle before the fort, but if the fighting went badly for the defenders, they quickly retreated inside the palisade again. Long sieges seldom succeeded, because most forts were equipped with reserves of food and an adequate supply of water. The defenders remained ready to repulse assaults with their bows and waited until their opponents tired of the siege.[19]

In warfare, as in hunting, an Indian's bow was his most valuable weapon. Martin Pring described the bows he saw at Plymouth

Harbor in 1603 as "five or six foot long of witch hazel, painted black and yellow, the strings of three twists of sinews, bigger than our bow-strings." A surviving bow taken from an Indian in Sudbury in 1660 is 66 3/4 inches long and was made from a single stave of hickory. The limbs are wide with a flat belly and a convex face. The grip is narrow and rectangular in cross section but comfortable to hold. An exact replica proved to have a "pull" of forty-six pounds at full draw and was capable of a maximum range of 173 yards with a special lightweight flight arrow.[20]

Compared with English long-bows, which usually pulled between sixty and seventy-five pounds, the New England bows were not powerful, but at moderate ranges they could easily kill a man. William Wood concluded that the Indian weapons were "quick, but not very strong, not killing above six or seven score [a maximum of 140 paces]." James Rosier, who drew an Indian witch hazel bow during a voyage along the Maine coast in 1605, said that the weapon was only powerful enough "to carry an arrow five or six score strongly."[21]

The deadly potential of Indian bows was demonstrated to a number of unlucky explorers and colonists. One of the Frenchmen killed by Indians at Chatham Harbor in 1606 was pierced by an arrow which also pinned his small dog to his body. Thirty years later, the Pequots slew several of Lion Gardiner's men at Fort Saybrook. Gardiner found "the body of one man shot through, the arrow going in at the right side, the head sticking fast, half through a rib on the left side." He sent both the arrow and the man's rib to officials in the Massachusetts Bay Colony, "because they had said the arrows of the Indians were of no force."[22]

Every male Indian learned to make or repair arrows, and most could produce a decent bow. Craft specialization was also practiced, however. Roger Williams said that the Narragansetts "have some who follow only making of bows, some arrows." These craftsmen were able to select the best materials for arrow shafts, projectile points, and bows. Their superior technological skills earned them respect and made their products more valuable than the work of average Indians. During an intertribal war, a cracked bow or warped arrow could be a very serious liability; most of the archery equipment used in combat was probably made by specialists in each tribe.[23]

Arrows made by New England Indians received the praise of

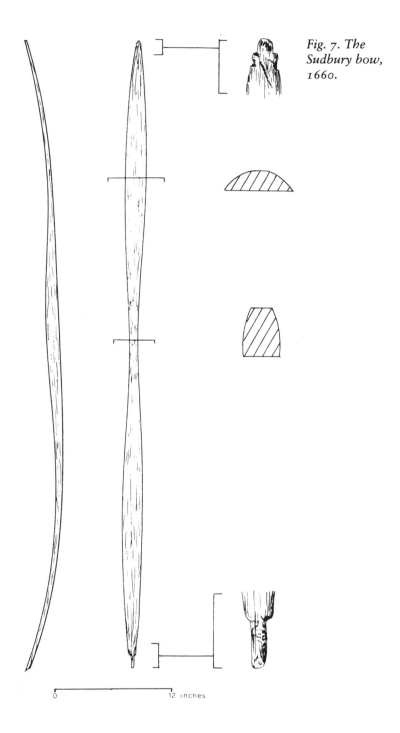

Fig. 7. The
Sudbury bow,
1660.

0   12 inches

European visitors, who were familiar with archery equipment. Martin Pring examined arrows which were "a yard and a handful long not made of reeds, but of a fine light wood very smooth and round with three long and deep black feathers of some eagle, vulture, or kite, as closely fastened with some binding matter as any fletcher of ours can glue them on." Other sources identify the wood used for arrows as elder. In some cases, the Indians attached their projectile point to a foreshaft six to eight inches long which they inserted in the end of the arrow shaft. This practice, useful in hunting and in warfare, allowed the point to separate easily from the long shaft after striking the target. William Wood explained that the arrow would "be left behind for their finding and the pile only remain to gall the wounded beast."[24]

Arrowheads used in hunting big game and in combat were designed to penetrate deeply, cut severely, and remain lodged in the victim. An arrow has little shocking power on impact, but with a sharp head it becomes a deadly, hemorrhage-producing projectile. Points of dense stone were widely used in New England before extensive contact with white men but seem to have gone out of favor by the early seventeenth century. In 1524, Verrazano observed that the Indians around Narragansett Bay used "sharp stones" instead of the iron arrowheads with which he was familiar. Later visitors recognized points made of bone, antler, native copper, and even eagle talons and horseshoe crab tails.[25]

The stone points that had played such an important role in aboriginal warfare before the establishment of English colonies were predominantly large triangular arrowheads without side notches or stems. Most were made of felsite, quartz, or quartzite found in the region. Along the coast, Indians often gathered beach cobbles and used a hammerstone to produce rough pieces. To make a finished point, an Indian pressed an antler tool against a roughly shaped piece of stone and chipped off small flakes until he had

Fig. 8. An Indian hunting with a bow.

a thin triangle with sharp edges. Manufacturing an arrowhead from a piece of proper stone took less than half an hour, but in some cases a tribe had to quarry the material for their projectile points and transport it many miles before the final flaking was done. If necessary any warrior could quickly find stone to fashion into a few crude arrowheads, but the production of thousands of carefully-shaped points required good knowledge of mineralogy and of the techniques of pressure flaking dense rock. Craft specialization must have been involved in the production of stone arrowheads by New England tribes.[26]

Copper arrowheads similar in shape to the triangular stone types were common in New England despite a scarcity of native copper deposits in the region. Tribes probably imported most of the metal from plentiful sources around Lake Superior or from deposits in Nova Scotia. Networks of intertribal trade covered great distances in North America long before the introduction of European trade goods. In 1602, John Brereton asked an Indian on an island off Cape Cod to explain where his tribe got such a "great store of copper" for ornaments, utensils, and arrowheads. The Indian took "a piece of copper in his hand, made a hole with his finger in the ground, and withal pointed to the maine from whence they came." Brereton marveled at the many items the Indians hammered from copper and noted that "they head some of their arrows herewith much like our broad arrow heads, very workmanly made." These triangular points were probably best for military purposes, but Indians also made some copper arrowheads by wrapping a small piece of hammered sheet copper into a conical shape that would fit tightly over the end of an arrow shaft. One conical specimen, fashioned from the European brass that the Indians later adopted in place of native copper, was found imbedded in the spine of an Indian skeleton excavated in Massachusetts.[27]

Bone and antler were readily available to Indians for manufacture into several types of projectile points. The sophisticated points used to impale fish were usually made of bone, as were many hunting and military arrowheads. Although decay has greatly reduced the archaeological evidence on the use of bone and antler in New England, the accounts of explorers and early settlers of the region prove the popularity of arrowheads made from these materials.[28]

An Indian seldom went anywhere without his bow and a quiver full of arrows. Pring described the quivers he saw near Plymouth as "full a yard long, made of long dried rushes wrought about two handfulls broad above, and one handfull beneath with pretty works and compartments, diamond wise of red and other colors." Each tribe may have had a different type of quiver, but all were light, useful devices which allowed an Indian to move rapidly through the forest while carrying an adequate supply of ammunition. Thomas Morton explained how the red men, "with their bow in their left hand, and their quiver of arrows at their back, hanging on their left shoulder with the lower end of it in their right hand . . . will run away a dog trot until they come to their journey end."[29]

In combat a warrior could fire one arrow after another in rapid succession; the quiver at his back kept his ammunition in easy reach. Edward Johnson described Myles Standish's first skirmish with Indians, during which the English captain wounded an opponent with his firearm. He hit the Indian's right arm, "it being bent over his shoulder to reach an arrow forth his quiver, as their manner is to draw them forth in fight."[30]

Indians shot very well at short range, but were far less accurate with their arrows beyond fifty yards. Except when fighting in an open field, they tried to get relatively close to their opponents before using their bows. In hunting or forest warfare the ability to strike a fatal wound with the first arrow was a great asset. An arrow shot from a long distance could easily miss its target, hit overhanging branches in its arching flight, or be seen in the air and avoided.[31] When an archer shot at close range, his shaft flew so quickly and directly at the intended victim that the man or animal had little time to react.

Early colonists testified to the Indians' remarkable skills with the bow at reasonable distances. William Wood described the local tribesmen as "most desperate marksmen for a point blank object." Thomas Lechford concurred, saying that they were "very good at a short mark." This skill was the result of years of practice. Wood observed that Indians were "trained up to their bows even from their childhood" and claimed that "little boys with bows made of little sticks could smite down a piece of tobacco pipe every shot a good way off." He was particularly impressed with the Indians' ability to shoot quickly and hit difficult, moving targets. They could, in his words, "smite the swift running hind [female deer] and nimble winked pigeon without a standing pause or left eyed blinking."[32]

When Indians closed with their enemies, their most effective weapons were clubs or axes. A fight that began with exchanges of arrows often ended with a rush by axe- or club-swinging warriors. A well-crafted axe was also useful in such nonviolent but militarily-valuable functions as felling trees, building defensive palisades, or shaping wooden canoes. Axe heads were usually made from hard, igneous rocks common in New England. An Indian chose a stone near the size and shape he wanted for his axe. If he intended to use the axe for cutting down large trees, he would pick a stone larger than the sizes used in war axes and small tools. He seldom shaped an axe head by chipping flakes from it as he did when making an arrowhead. Instead he pecked and abraded the surface of the stone until the head was properly formed.[33]

The southern New England Indians used smooth, grooveless heads of various sizes in most of their war axes. Daniel Gookin said that warriors often carried "tomahawks, made of wood like a pole axe, with a sharpened stone fastened therein." He referred to the type of axe made by forcing the tapered end of a smooth stone head through a hole in a solid wood handle. This method of hafting was strong enough to withstand the rigors of combat and of light woodcutting as well.[34]

Indians also made axes with grooved stone heads for heavier work. They pecked a groove around the head where the handle would be fastened and then hafted the stone in one of several ways that made the axe capable of repeated hard blows. The most important military application for axes of this type was in the construction of fortifications, but such durable cutting tools contributed directly or indirectly to the Indian military system in many ways.[35]

Many Indians preferred clubs over axes for close fighting. Striking their blows at an opponent's head, they could kill a man with any of the types of clubs used in New England. Individual preferences probably determined the type of hand weapon carried by most warriors, but some tribes may have used only a certain type of club or war axe. Because early colonists used the Indian term "tomahawk" when referring to ball-headed wooden clubs, stone hammers, stone axes, or other striking instruments, their accounts often fail to define the exact weapon used by the Indians they observed. William Wood was more specific when he described the tomahawks carried by invading Mohawks as "staves of two foot and a half long, and a knob at one end as round and big as a football." John Josselyn confirmed that many New England Indians

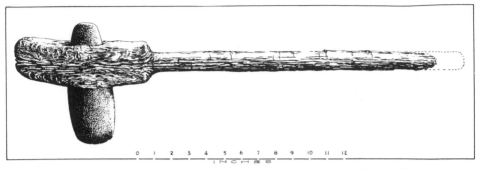

Fig. 9. Axe with smooth stone head set in original haft. This matches some early descriptions of "tomahawks."

also had these clubs, and the Reverend Increase Mather mentioned another form of war club, a "hammer" with a stone head. This hammer, which Mather said was a weapon of the Pequots, probably had a grooved head without a sharp edge and would therefore be similar to some of the Indians' pounding tools.[36]

Another close-range weapon, the spear, may have served as a symbol of rank among the Indians. Wood said that "their captains have long spears," but he did not mention their use by ordinary warriors. Some spears were headed with sharpened bone, and archaeological evidence also suggests the use of chipped stone points similar to cutting knives or large arrowheads.[37]

The men who planned the strategy of wars and commanded the military operations were from the ranks of the elite warriors whom the Wampanoags called pnieses. Sachems were primarily civil leaders and usually deferred to their most able military men in the conduct of wars. William Morrell explained in poetry how sachems depended on their finest warriors:

> These heads are guarded with their stoutest men,
> By whose advice and skill, how, where and when,
> They enterprise all acts of consequences,
> Whether offensive or for safe defense.

Experienced war leaders organized raiding parties and used their own reputations to attract eager young men to join them. In battles they were the "captains" who kept the other warriors in control and inspired them by personal example.[38]

In warfare as in most important enterprises, the Indians sought

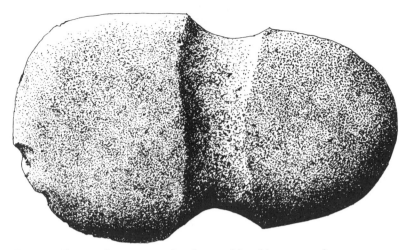

Fig. 10. *Grooved, stone axe head, capable of heavy work.*

the aid of supernatural powers and listened to the advice of priests called "powows." Few men in the tribe were more feared or respected than these mysterious priests, who claimed to be in contact with various deities and spirits. The warriors believed that they could be assured of victory if the powows performed the proper ceremonies and enlisted the support of major gods like Hobbamock and Kiehtan. Many of the dances, speeches, and ceremonies that preceded wars and individual battles were as religious in character as they were militaristic. Mary Rowlandson, a captive of the Indians in 1676, observed the elaborate ceremonial preparations for a raid on Sudbury. She thought the warriors "acted as if the devil had told them that they should gain the victory." They were so confident that "they went without any scruple, but that they should prosper."[39]

When Indians went to war, they took great pains to look as fierce and impressive as possible. William Wood described how they painted their faces with a "diversity of colours, some being all black as jet, some red, some half red and half black, some black and white, others spotted with divers kinds of colours, being all disguised to their enemies, to make them more terrible to their foes." Daniel Gookin wrote: "the men in their wars, do use turkey or eagle's feathers, stuck in their hair, as it is traced up in a roll. Others wear deer shuts, made in the fashion of a cock's comb dyed red, crossing their heads like a half moon." The men also deco-

Fig. 11. Nineteenth-century engraving of an imagined Indian religious ceremony. The artist purports to show powows, but gives little sense of the beauty and mysticism of Indian faith.

rated their bodies with valuable wampum and various types of jewelry. They were proud of their warlike appearance, and their attire probably helped them to feel confident and aggressive in battle.[40]

The painted and fiercely decorated combatants in an intertribal war were much better organized than their wild appearance might suggest. Although the warriors retained more individual freedom of action than did soldiers in the disciplined ranks of a European army, they did employ tactical plans and usually responded to the commands of their leaders. Their ambushes and raids required a high level of tactical skill and coordination. Just to move a body of men secretly through the forest was a serious military problem; fighting there presented difficulties that only good tactical control could overcome.[41]

Most English observers failed to see the tactical sophistication that often shaped aboriginal military actions. Men who fought apart from their comrades, who hid behind trees and fired at will, seemed by European standards to have no real military skill or tactical order.[42] But the formation and maneuvers of the parade ground were impossible in wooded terrain, and without dense formations, the volley fire favored by Europeans was ineffective. The tactics of the southern New England Indians fit their environment and the limited goals of their wars. Warriors executed complicated tactical operations in the forest and made the terrain and vegetation their allies. Striking by surprise and relying on the cover and concealment provided by the forest, they often won quick victories with few or no losses.[43]

The ambush was the Indians' favorite form of offensive action. They expended much effort trying to surprise enemies moving through the forest. Waiting in hidden positions along well-traveled trails, they struck without warning when their victims passed by. If scouts were able to sight an enemy group and determine its route of movement, a war party could set an impromptu ambush in its path. Sometimes Indians even lured opponents into prearranged traps. They borrowed this ambush technique from hunting practices, but instead of luring their quarry with food or calls, they sent out human decoys to flee from a larger force and lead it into a trap.[44]

When the Indians set an ambush, they tried to pick a location where they could achieve complete surprise and hold their opponents long enough to inflict casualties or take prisoners. The choice of terrain was very important, for natural features like cliffs, rivers, or lakes could block a route of escape and make it easier to pin the victims in a vulnerable position. If there were no obstacles on at least one side of the location, the Indians would usually try to close a circle around the enemy at the moment of attack. The ambush position also had to offer good concealment for the attacking force; catching your opponents unaware was the key to success.[45]

In raids, as in ambushes, surprise and tactical coordination were essential. War parties conducted raids against stationary objectives such as villages and camps and sometimes traveled great distances to launch their attacks. They sent scouts ahead of the main party to reconnoiter. Hunters might seek game enroute to the enemy territory but were careful to stay behind the rest of the advancing force. Taking care to avoid premature discovery, the warriors often relied on darkness to shield their movements as they neared their objective.[46]

A war party might wait until it was near a village or an enemy

Fig. 12. Warriors with bows.

force before making final tactical plans. Although little is known of how the southern New England Indians planned their military actions, the writings of Samuel de Champlain provide a good description of northern Algonquian and Huron practices. The use of tactical planning and rehearsal was probably common throughout the woodlands of the northeast.

Having learned from their wizards [powows] what is to happen to them, the chiefs take sticks a foot long, one for each man, and indicate by others somewhat longer, their leaders, then they go into the wood, and level off a place five or six feet square, where the head man, as sergeant-major, arranges all these sticks as to him seems best. Then he calls all his companions, who approach fully armed, and he shows them the rank and order which they are to observe when they fight with the enemy. This all these Indians regard attentively, and notice the figure made with these sticks by their chief. And afterwards they retire from that place and begin to arrange themselves in the order in which they have seen these sticks. Then they mix themselves up and again put themselves in proper order, repeating this two or three times and go back to their camp, without any need of a sergeant to make them keep their ranks, which they are quite able to maintain without getting into confusion. Such is the method they observe on the warpath.[47]

In 1637, the Narragansetts gave the English some good advice on how to conduct a raid on a specific, unfortified camp of the Pequots. In a letter to Governor John Winthrop, Roger Williams reported the following Indian suggestions: "That the assault would be in the night, when they are commonly more secure and at home, by which advantage the English, being armed, may enter the houses and do what execution they please" and "That before the assault be given, an ambush be laid behind them, between them and the swamp, to prevent their flight, etc." In all the years since 1637 no one has really improved on this plan for a surprise attack with a blocking force set in the ambush.[48]

Indians timed their raids to catch their opponents off guard and to confuse them. Attacks in darkness, at first light, and during storms or conditions of heavy fog gave the raiders a better chance of success. Some of the powows claimed to have power over weather and tried to call up storms for military advantage. Edward Winslow said that they took "advantage of their enemies in their houses" when wild weather kept people inside. "At such times they perform their greatest exploits, and in such seasons, when they are at enmity with any, they keep more careful watch than at other times."[49]

The tactics and weaponry of the southern New England Indians could have produced heavy loss of life in an unrestricted form of warfare. Yet the military actions of these warriors seldom resulted in heavy casualties on either side. They had no desire to destroy an opposing force completely or to wipe out an enemy community. Early colonists, familiar with the slaughter of seventeenth-century continental wars were amazed to see the Indians, who wore no armor and carried no shields, fight for prolonged periods with minimal losses. When compared with the Thirty Years War in Europe, Indian wars looked like games to English observers, who failed to recognize the serious intent of this different type of combat.[50]

Colonists with military experience were scornful of the limited wars of the Indians. Captain John Underhill of Massachusetts Bay, a former professional soldier, wrote of a battle between the Pequots and the Narragansetts that their combat seemed "more for pastime than to conquer and subdue enemies." He sarcastically concluded that "they might fight seven years and not kill seven men."[51]

Underhill was particularly critical of battles fought in fields, where the warriors stood at considerable distance and shot their arrows in high arcs, despite the ease of dodging such indirect fire. This combat was, of course, the form of Indian warfare which colonists were most likely to observe and the form which was most similar to action on European battlefields. William Wood, a civilian who reported on the culture of the New England Indians, shared Underhill's scorn for the open battles. The Indians failed to fight in precise formations or to hold their ground with the tenacity that the English expected of soldiers:

> Being thus armed with their warlike paint, the antique warriors make towards their enemies in a disordered manner, without any soldier-like marching or warlike posture, being deaf to any word of command, ignorant of falling off, or falling on, of doubling ranks or files, but let fly their winged shaftments, without either fear or wit; their artillery being spent, he that hath no arms to fight, finds legs to run away.[52]

Even when warriors rushed each other and fought at close range, they did not kill enough men to meet some colonists' standards of serious combat. John Mason, a European veteran like Underhill, was disappointed with the results of a battle between his Mohegan allies and a band of Pequots. The Mohegans, who had been under the sway of the Pequots, broke away by 1634 and joined the English against their former associates in 1637. A hundred Mohegans ran at about sixty Pequots "and met them and fell on pell mell striking and cutting with bows, hatchets, knives, etc. after their feeble manner; indeed it did hardly deserve the name of fighting."[53]

Roger Williams, a friend of the Indians, admitted that intertribal wars were "far less bloody, and devouring than the cruel wars of Europe," but unlike many of his contemporaries, he saw no reason to scorn the Native Americans for this. Estimating that there were "seldom twenty slain" in even the largest battles, he went on to explain how Indians used trees as shields in the forest and dodged arrows in the open. His respect for the courage of Indian warriors is obvious in this description of combat:

> When they fight in a plain, they fight with leaping and dancing, that seldom an arrow hits, and when a man is wounded, unless he that shot follows upon the wounded, they soon retire and save the wounded: and yet having no swords, nor guns, all that are slain are commonly slain with great valour and courage: for the conqueror ventures into the thickest, and brings away the head of his enemy. [54]

Most colonists in the first three quarters of the seventeenth century thought of warfare in terms of formal battles and single-minded dedication to the destruction of the enemy. They wanted the Indians to fight in the open, but with more discipline and with greater willingness to suffer and inflict grievous losses. They did not understand the Indians' attitude toward warfare and did not realize the sophistication and military effectiveness of their ambushes and raids. These Europeans judged the military skills of the Indians primarily from reports or observations of a few open battles and condemned the more clandestine tactics as devious "skulking" instead of real warfare. [55]

The wars between southern New England tribes and the Mohawks, who lived to the west beyond the Hudson River, were particularly upsetting to colonists, because little fighting was done in open engagements. John Josselyn said, "Their fights are by ambushments and surprises, coming upon one another unawares." When five Mohawks were detained in Cambridge in 1665, the colonists tried to explain to them how a proper war should be conducted. Daniel Gookin reported that the Mohawks were told to "fight with their enemies in a plain field." The colonists chastised them for acting "more like wolves than men" and for fighting "in a secret, skulking manner, lying in ambushment, thickets, and swamps by the way side, and so killing people in a base and ignoble manner." [56]

The European culture did have a powerful influence on Indian patterns of warfare, but not always in ways that the colonists intended. No one wanted more wars between tribes or between the Indians and the English; the colonists simply wanted the Indians to conform to European standards of military conduct. European technology adopted by the Indians or modified to fit their particular needs increased the potential for serious bloodshed in forest combat. Killing became simpler with improved weapons, and methods of warfare were adapted for the deadlier technology. The Indians retained their high regard for secrecy and surprise in warfare, but they learned from the English just how terrible war could be. Colonists introduced the Indians to new technology and then, unwittingly, inspired them to use their improved military capability in a form of total warfare that was outside their aboriginal experience.

CHAPTER II

The Arrival
of the White Man

The coming of the white man with his devastating diseases, his hunger for land, and his military system ultimately destroyed the culture of the southern New England tribes. Yet many Indians initially welcomed the Europeans and sought the products of their technology. Articles of European manufacture enhanced the status of a warrior, made his labor and hunting easier, and ominously improved his ability to kill other men in combat. The Indian military system and Algonquian culture in general began to change significantly in the early seventeenth century as a result of increasing contact with the white man. This process was accelerated with the expansion of European trade and the formation of colonies in the northeast. Many of the objects, practices and ideas of European civilization were adopted by Indians during this period.

Until the early seventeenth century, very few white men visited the shores of southern New England. Giovanni da Verrazano had the first recorded meeting with the Indians of that region in 1524, but his vessel stayed only a short time, and the few items he traded with the inhabitants had little effect on their culture. The men of many nations who fished the waters off Newfoundland during the sixteenth century rarely sailed as far south as New England. As the seventeenth century began, however, some fishermen and entrepreneurs became interested in the possibility of good fishing and fur trading in the New England area. Ships were soon exploring in earnest the coasts of Maine and Massachusetts.[1]

Between 1602 and 1605, voyages by Bartholomew Gosnold, Martin Pring, and George Waymouth provided convincing evidence that the waters from Maine to Cape Cod were excellent for fishing. The Maine coast in particular became the destination for many fishermen and traders, but a number of ships also visited southern New England in the two decades before the Pilgrims came. White men entered Plymouth Harbor at least six times before 1620.[2]

Early explorers and fishermen gave the Indians items of European manufacture in exchange for furs, souvenirs, or provisions. When Gosnold arrived on the Maine coast in 1602, he found that other Europeans had already given one group of Native Americans a variety of articles. John Brereton described the incident in his narrative of Gosnold's voyage: "eight Indians, in a Basque-shallop with mast and sail, an iron grapple and a kettle of copper, came boldly aboard us, one of them apparelled with a waistcoat and breeches of black serge, made after our sea fashion, hose and shoes on his feet; all the rest (saving one that had a pair of breeches of blue cloth) were naked." Gosnold and his men also visited southern New England and traded with some of the Indians of the Cape Cod area. Brereton reported that the English supplied several with "certain trifles, as knives, points, and such like which they much esteemed." Gabriel Archer, another member of the expedition, said that Gosnold gave an Indian "a straw hat and a pair of knives; the hat awhile he wore, but the knives he beheld with great marveling, being very bright and sharp; this our courtesy made them all in love with us."[3]

The New England Indians were usually glad to trade with white men, but not all the early contact between Europeans and Indians was pleasant. Some unscrupulous traders cheated the Indians, stole from them, and even kidnapped a number. Indians, often because of prior provocation, also committed thefts and were sometimes hostile to Europeans. Occasional murders, ambushes, and raids made exploration and trade dangerous in certain areas, but the firearms of the white men soon proved to be powerful deterrents. Most Indians learned to respect European weapons and to exercise caution in their dealings with armed traders. Also, the Indians' desire for products of European technology was strong enough to promote considerable cooperation.[4]

The Pilgrims and other early colonists settled in southern New England with little opposition, not only because the Indians feared their muskets and wanted European trade goods, but also because of the effects of a disastrous plague which had struck many tribes a few years before the Pilgrims arrived. About 1616, a European visitor brought a disease which has never been conclusively identi-

Fig. 13. Landing of the Pilgrims.

fied to the coast of New England, and within a few years the entire aboriginal population of over 75,000 had been reduced by almost half. The coastal tribes from Cape Cod to the Penobscot River in Maine were affected most seriously. When Thomas Dermer sailed south from Maine in 1619, he reported: "I passed alongst the coast where I found some ancient plantations, not long since populous now utterly void; in other places a remnant remains, but not free of sickness." John Josselyn said that the disease caused "a great mortality amongst them, especially where the English afterwards planted."[5]

This plague, which struck some tribes terribly and others slightly or not at all, disrupted the balance of power and increased inter-tribal aggression in New England. The disease depopulated all of the Wampanoag villages near Plymouth Harbor. Robert Cushman reported: "We found the place where we live empty, the people being all dead and gone away." When in 1621 a party from Plymouth traveled west to the village of Massasoit, the principal sachem of the Wampanoags, they found "the people not many, being dead and abundantly wasted in the late great mortality." The Narragansetts, however, "lived but on the other side of that great bay, and were a strong people and many in number, living compact together, and had not been at all touched with this wasting plague." Fearing their now threatening Indian neighbors, the decimated Wampanoags chose to aid the Pilgrims and thereby gain a powerful ally. Farther north the

Fig. 14. Nineteenth-century engraving of the "Great Mortality" among the New England Indians caused by the transmission of an unknown European disease in approximately 1616.

few hundred survivors of the once-mighty Massachusetts were happy to place themselves under the protection of the Puritans, who arrived in 1629.[6]

European diseases, particularly smallpox, continued to take a heavy toll of the Indians long after the great plague of 1616–1617. The Narragansetts and tribes in Connecticut, who had avoided the first onslaught of imported illness, did not escape the widespread epidemics of smallpox in the 1630s. Deaths of tribal leaders and the sudden weakening of some groups in comparison to others added more instability to the tense environment in southern New England.[7]

The dangers of contact with Europeans did not discourage the Indians from seeking more trade with them. The demand for trade goods grew stronger as the Indians became accustomed to and even dependent on the new articles. Not only did an increasing percentage of the Indian population use European products but also, because of Algonquian burial customs, many valuable weapons, tools, and other items were interred with the dead.[8]

Although Indians welcomed many of the articles offered as gifts or as trade items, they were selective in their adoption of foreign products. Favored goods usually satisfied functional or symbolic needs already existing in the aboriginal culture. A particular object might be more durable, efficient, or attractive than an Indian artifact serving a similar purpose. The function and meaning of an artifact could, however, change dramatically as it passed from one culture to another. Indians flatly rejected a number of European items, made physical modifications to others before adopting them, and acquired some simply to make use of the raw material contained in them. Native American craftsmen acknowledged the value of imported metals but retained their respect for traditional forms and ornamentation when they cut up brass or iron goods to create products that were distinctly Indian in appearance.[9]

Even the form of a European article was often familiar to its new owner. An Indian could easily recognize a trade hatchet as an improved version of his stone-headed axe or tomahawk. He was perhaps surprised by the toughness of the hatchet's wrought iron head and the cutting ability of its sharpened steel edge, but metal artifacts of native copper had been made for centuries in North America. The warrior needed no lessons in the military applications of the hatchet.[10]

Certain European weapons and tools soon became prized possessions of New England Indians. The Indians used their trade hatchets in close combat, in the construction of fortifications and in countless routine tasks. For military use, they much preferred the versatile hatchet over the sword carried by European soldiers and most colonial militiamen. However, Indian craftsmen did convert sword blades into effective spear points. They also recycled broken or surplus hatchet heads to make chisels, wedges, and other small tools. Knives, drills, hammers and many other utilitarian artifacts helped to change the technological capabilities of Indians and to influence their military activities.[11]

No European artifact adopted by the Indians had a more dramatic effect on their military system than did the firearm. Despite the superior rate of fire of aboriginal bows, Indians quickly recognized important advantages in the use of muskets and carbines both on the hunt and on the warpath. A firearm did not resemble a bow as an iron hatchet resembled a stone hatchet, but the function of the gun was the same as that of the bow—to kill at a distance. Warriors

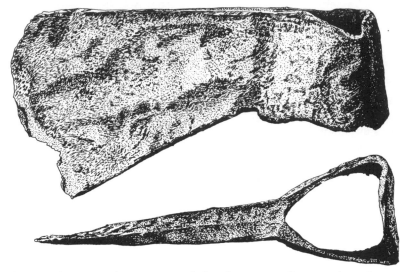

Fig. 15. *Seventeenth-century trade hatchet, or small axe. Side and top views of wrought iron head with steel edge insert.*

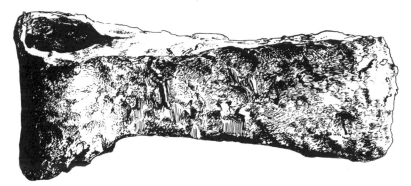

Fig. 16. *Axe head from a seventeenth-century Wampanoag burial in Rhode Island, at Burr's Hill.*

Fig. 17. *Knife blade with tang, from Burr's Hill.*

experienced in the use of bows saw that bullets flew much faster than arrows and took a more direct route to the target. The heavy lead projectiles were less susceptible to deflection by overhanging branches or light brush, almost impossible to dodge, and more damaging upon impact. Bullets would even penetrate the European armor worn by some colonists. Firearms could also be loaded to fire a number of small balls at one shot, making it even easier to strike an opponent or game animal.[12]

The first reaction of Indians to firearms was shock and awe. The noise, the flash of exploding powder, and the cloud of smoke must have been frightening to even the bravest warrior. James Rosier reported that when Indians on the Maine coast witnessed the firing of English weapons in 1605, they were "most fearful, and would fall flat down at the report of them." Roger Williams said that the Narragansetts recognized a "consimilitude between our guns and thunder, they call[ed] a gun 'peskunck' and to discharge 'peskhommin,' that is to thunder." By relating the strange weapon to something which they knew, the Indians not only acknowledged its power, but also prepared for its acceptance into their culture.[13]

William Bradford of Plymouth witnessed the Wampanoags' sudden surge of desire for firearms in the 1620s. After these Indians "saw the execution that a piece [a musket] would do, and the benefit that might come by the same, they became mad (as it were) after them and would not stick to give any price they could attain to for them; accounting their bows and arrows but baubles in comparison of them."[14] Yet, even in their rush to acquire the white man's guns, Indians showed excellent judgment in assessing the relative values of the several types of firearms which Europeans used in America.

The vast majority of firearms carried by early explorers and by the Pilgrims, who came to New England in 1620, were muskets called "matchlocks" after their firing mechanism. The well-organized Puritans of Massachusetts Bay, who arrived almost a decade later, were also armed primarily with matchlock muskets, although most of the muskets owned by their company were more advanced, self-igniting flintlocks. During the first half of the century, colonists relied heavily on matchlocks for their military defense.[15]

Matchlock muskets, standard equipment in European armies, were relatively inexpensive for prospective colonists. Simple in operation, the lock lowered a lighted match, held in a device called a serpentine, into an open pan of priming powder. By pulling a trigger on some weapons or depressing a lever on others, the musketeer forced the serpentine to rotate against a restraining spring, thus bringing the match into contact with the priming powder and setting off an explosive train leading from the pan through a touch hole to the propellant charge in the barrel. The projectile, usually a large lead ball weighing a twelfth of a pound, was sent on its way with great force.[16]

European armies found the matchlock musket an effective arm for massed formations. Although it weighed up to twenty pounds, was inaccurate beyond fifty yards, and had to be fired using a forked rest, it performed well during European infantry actions in which ranks of musketeers fired concentrated volleys at close range. However, a weapon suitable for the battlefields of Europe was not necessarily adequate for warfare in the forests of New England.[17]

In Europe, soldiers fought opponents who were willing to accept battle under mutually advantageous situations and to forego actions in bad weather, darkness, or forested terrain. The militiamen of New England in the seventeenth century faced Indian warriors with a long tradition of success through stealth and surprise. Indians used the forest as an ally against their enemies. They attacked when and

Fig. 18. Operation of a matchlock musket.

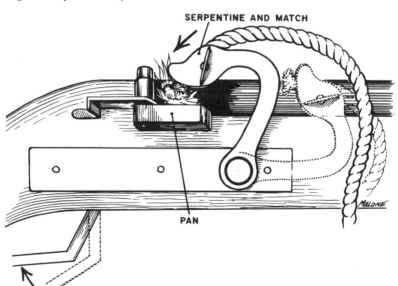

SERPENTINE AND MATCH

PAN

where they chose, striking when least expected and taking advantage of every weakness an enemy revealed.

Many characteristics of the matchlock proved to be liabilities in the New World. A musketeer had to light his match, a cord treated with saltpeter or gunpowder, in advance of any action. Failure to have a ready match could prove fatal, because attackers were not likely to give their enemies a chance to start a fire. A musketeer in a combat situation was expected to keep both ends of his match lighted, to adjust it frequently as it burned down, and to blow the ashes off the smoldering tip which was clamped in the serpentine of his musket. Since a match burned at a rate of up to nine inches an hour, considerable quantities of the special cord had to be carried into the field and kept in good condition.[18]

In rainy weather a musketeer tried to keep his match dry in his hat or under cover in some other way. The effort was troublesome and the results often futile. Rain not only extinguished burning match tips, but also could ruin one's spare match. Pilgrim musketeers on Cape Cod had to trim their matches after a rainy night in 1620, for upon checking their muskets in the morning, they found that "few of them would go off because of the wet."[19]

As early as 1607, Indians recognized weaknesses in the ignition system of the matchlock musket. William Strachey, in his narrative of the voyage of Gilbert and Popham to Sagadahoc, explained that a group of Indians "subtilely devised how they might put out the fire in the shallop [boat], by which means they saw they should be free from the danger of our men's pieces, and to perform the same, one of the savages came into the shallop and taking the fire brand, which one of our company held in his hand there to light the matches, as if he would light a pipe of tobacco, as soon as he [the Indian] had gotten it into his hand he presently threw it into the water and leapt out of the shallop." Although these Indians had discovered the Achilles' heel of the matchlock, they were unwilling to bet their lives on their solution. When Gilbert ordered his men "to present their pieces" in a desperate bluff, the shaken warriors decided not to risk the possibility that the muskets could still fire without lighted matches. They took their bows and fled into the forest.[20]

The flaws in the matchlock became more obvious to Indians as they tested weapons acquired in trade and as they witnessed the problems experienced by musketeers in some of the early skirmishes between white men and Indians. A warrior who suddenly darted

Fig. 19. Pilgrims in their first military action against Indians, on Cape Cod in 1620.

from the brush was no easy target for a European whose weapon required a separate rest to support it in firing. Even worse, the musketeer might be surprised when he was unable to fire his matchlock at all because of an extinguished match or wet priming powder. His chances of ambushing the Indians in their own forests were minimal with a lighted match that not only glowed in shadows or darkness, but also gave off a recognizable odor. Even if he managed to fire his weapon, the process of reloading it was slow, complicated (over forty separate motions were prescribed in most military manuals), and dangerous.[21]

Captain Myles Standish, an experienced professional soldier and an influential military leader in Plymouth Colony, preferred a type of firearm more advanced than the matchlock. He brought with him to New England a "snaphaunce," which was self-igniting and required no rest to steady it while firing. One advantage of this weapon was shown during an attack by Indians on Cape Cod in 1620; Standish was able to fire while some of his men were calling "for a firebrand to light their matches."[22]

In the seventeenth century all weapons whose actions operated by striking flint against steel to produce an igniting spark were called

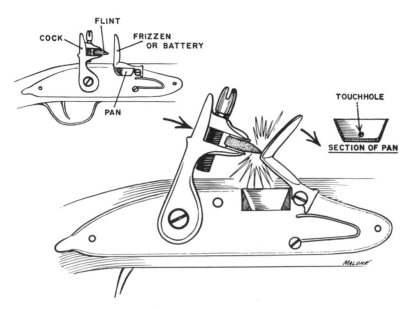

FLINT

COCK

FRIZZEN
OR BATTERY

PAN

TOUCHHOLE

SECTION OF PAN

Fig. 20. Operation of a flintlock musket.

"snaphaunces" or "firelocks." Modern experts distinguish many varieties of flint arms. The true snaphaunce, rarely used in New England, is different from other flintlocks primarily because the cover to the priming pan is separate from the steel "battery" against which the flint strikes. In all flintlocks the flint is clamped in a cock, and, in all but a true snaphaunce, the battery and pan cover are forged in one piece, so that the striking action of the cock against the battery not only produces sparks, but also opens the pan into which the sparks fall. Other minor differences in the mechanisms of flintlocks have resulted in modern distinctions between the English lock, dog lock, and "true" flintlock, all used in New England in the seventeenth century. Because no such distinctions were made in the sources, it is best for the purposes of this study to use the term flintlock to refer to any flint arm. The sources quoted here may use the terms snaphaunce or firelock to mean any member of the general class of flint-ignited firearms.[23]

The flintlock, like the less expensive matchlock, was loaded with powder and ball through the muzzle; the charge was rammed down with a rod; and the priming pan was filled with fine, sensitive powder. Because of the absence of a lighted match, the process of loading was simpler, quicker, and far less dangerous. The flintlock was also faster and easier to fire than the matchlock. A slight movement of the trigger released the spring-powered cock and produced rapid ignition of the priming powder. When a musketeer fired a matchlock, he had to move the triggering level significantly and with some care, because the serpentine, holding the delicate, lighted tip of the match, was rotated into the priming pan by the mechanical movement of the lever.[24]

Off-hand shooting at moving targets was possible because of the rapid ignition produced by the flintlock. A musketeer needed no rest; he could "snap-shoot" his weapon at a suddenly appearing enemy with some hope of success. He no longer needed to light a match in advance of any action. The flintlock even made it possible for a colonist to hide in ambush with a ready musket that gave no hint of his presence to the enemy. The flintlock was a much better weapon for forest warfare than the matchlock, but the Indian was quicker to realize this than the average colonist.

Despite increasing evidence that the matchlock was unsuited for forest warfare, colonial governments were slow in requiring militiamen to equip themselves with expensive flintlocks. The fact that European armies still fought with matchlocks had a powerful influence. The best musketeers of the day fired volleys on command, with little regard for aiming and with justified confidence in the capability of a lighted match to ignite dry powder. Formal battles in nice weather on open fields made the cumbersome matchlock appear effective; military commanders in England did not worry about ambushes, night attacks, or enemies who took cover behind trees.[25]

Standish, John Endicott and other perceptive leaders in the New England militia units recognized that a different environment and opponents from another culture made some changes in equipment necessary. They convinced their colonial governments to purchase flintlocks as "common arms" owned by the colony. They also urged individual colonists to spend the extra money required for flintlocks when they armed themselves for militia service. The transition to widespread use of the flintlock in the colonial militias took time, however.[26]

During the Pequot War in 1637, the first major conflict between New England colonists and Indians, the forces of Massachusetts Bay and Connecticut won easily over a tribe that had very few firearms.

John Underhill revealed the mix of muskets used by the Englishmen when he described a volley involving "both flint and match."[27] The stunning victory, gained with the aid of Indian allies and with the use of torches against a palisaded village, taught the colonists nothing about the military potential of the Indians and must have left many thinking that matchlocks were good enough for such weak and ill-equipped opponents.

Plymouth Colony, small and in close contact with the powerful Narragansetts, moved more quickly than its neighboring colonies to limit the use of matchlocks. When trouble with Indians caused the New England colonies to mobilize field forces in 1645, Governor William Bradford noted proudly that the forty men sent by Plymouth "were well armed all, with snaphaunce pieces." A year later the colony required each town to maintain in reserve, as public arms, two flintlocks for every thirty men. The transition to flintlocks was apparently complete in the colony's militia bands years before the outbreak of King Philip's War in 1675. The lessons learned in that devastating struggle prompted Plymouth officials to ban the military use of matchlocks in 1677.[28]

As late as 1650, the Connecticut general court was ordering towns to keep a supply of match in addition to their small quota of publicly owned flintlocks. Although the colonists who fought in King Philip's War were armed almost entirely with flintlocks, their governments had not yet formalized the transition in law. Massachusetts Bay Colony finally enacted such a law in 1693; Connecticut Colony never bothered.[29]

In sharp contrast to the majority of English colonists, New England Indians chose flintlocks over matchlocks almost immediately. They knew how to hunt and fight in the forested terrain of the eastern woodlands, and they knew at once that a weapon dependent on a lighted match did not compare with a self-igniting flintlock. The practices of trained European armies meant nothing to them; Indians simply chose the weapon best suited for their hunting and their military tactics.[30]

Selecting the best weapon for hunting or fighting may have been easy for the Indians, but acquiring flintlock muskets in large numbers was more difficult. Before the Indians could arm themselves effectively, they had to develop a trade economy of large scale and considerable sophistication. If they wanted European products, they had to have something to trade for them; if they were to develop a complex network of trade between tribes and colonies, they would even need an accepted medium of exchange.

The Indians of New England began to trade European goods among themselves even before all the tribes had contact with white men. The fur trade, stimulated by a soaring international demand for beaver in the early seventeenth century, was the source for most of the trade goods that were soon in great demand among all the Indians of the northeast. Through individual traders and monopolistic companies of several nations, some tribes acquired large quantities of European tools, utensils, ornaments, textiles, and weapons. Intertribal trade then spread these articles throughout New England.[31]

The French and the Dutch were more aggressive fur traders than the English during most of the early seventeenth century. French traders at posts in Acadia and on the St. Lawrence River exchanged vast amounts of European trade goods for beaver killed by the Indians. Dutch trade in the Hudson River Valley and along the coasts of southern New England developed more slowly than the French operations but was extensive by the third decade of the century. English companies on the Maine coast, roving fisherman-traders, and permanent English colonies added their trade goods to the increasing supply of European products that entered the network of intertribal trade.[32]

In the late 1620s, New England Indians adopted a standard medium of exchange called "wampum" or "wampumpeag," which was very important in their trade with each other and with the Europeans. Beads made from certain sea shells, primarily the whelk, the periwinkle, and the hardshell clam known as a "quahog," were strung together and measured by fathoms. Indians found these shells in greatest number along the shores that flanked Long Island Sound and Narragansett Bay. Until the Dutch in New Netherland began to use wampum in their trade and to assign it a monetary value, these shell beads, which could have sacred significance, had been worn as ceremonial jewelry, used in rituals of tribute and consolation by Indian leaders, and sometimes offered in exchange for other items. In 1627, the Dutch induced the Pilgrims of Plymouth to use wampum in trade, and by the 1630s it was legal tender in the New England colonies.[33]

Thomas Morton used his personal experience among the Indians

Fig. 21. Dutch traders with Indians. The early acceptance of wampum as a medium of exchange in New Netherland facilitated an extensive network of trade which extended into southern New England.

to explain why traders accepted wampum as willingly as they did furs: "We have used to sell them any of our commodities for this wampumpeag, because we know we can have beaver again of them for it: and these beads are current in all the parts of New England, from one end of the coast to the other."[34]

Tribes on the coast of southern New England, particularly the Narragansetts, began to produce wampum in quantity, making themselves wealthy by supplying it to inland tribes in exchange for furs and trade goods. William Wood said of the Narragansetts that "the northern, eastern, and western Indians fetch all their coin from these southern mintmasters." William Bradford explained that the manufacture of wampum by coastal tribes made them "rich and powerful and also proud thereby." It involved the tribes in extensive Indian networks of trade reaching outside New England and, at the same time, increased their dependence on the white men who supplied trade goods and treated wampum as a currency. Up to the

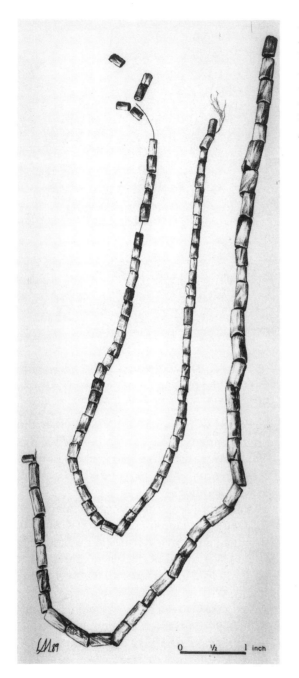

*Fig. 22. Two
strings of
wampum, from
Burr's Hill. Each
tubular bead was
laboriously
manufactured
from shell.*

0   ½   1 inch

mid 1660s, the Mohawks' continual interest and occasional military intervention in southern New England tribal affairs was largely the result of their need for a stable supply of wampum.[35]

European tools and weapons acquired in trade allowed Indian hunters to kill more beaver for the expanding market. There were no steel traps for the hunters to set, and the use of castoreum as a bait had not yet been discovered. Indians caught some beaver in "dead-fall" traps, which dropped a heavy weight on the unfortunate animals, and shot others from canoes or from the banks of streams and ponds. However, they got most of their beaver through back-breaking labor done with European tools. They broke beaver-constructed dams to drain ponds and isolate the animals' lodges. Then they chopped their way into the sturdy dwellings. They also cut through ice to get at lodges in the winter and to make holes so they could shoot escaping animals who sought access to air after fleeing their homes. They even dug beaver out of burrows near the water's edge. European axes, ice chisels, and digging tools were invaluable in these labors, but barbed iron arrowheads and muskets also aided in taking many of the amphibious and elusive creatures.[36]

Much of the European equipment which enabled the Indians to provide an ample supply of furs and meat for white men also caused more bloodshed in intertribal warfare and made many of the southern New England tribes potentially dangerous to colonial settlements. Despite many efforts by colonial authorities to prevent the sale of firearms to the Indians, the trade was too lucrative to be stopped and became a very serious problem in the years between the Pequot War and King Philip's War. However, firearms were only a part of the military equipment given to the Indians or made by them from European articles and scraps of metal. As mentioned above, the Indians used trade hatchets and knives as weapons and made spears by fixing pieces of sword blades to the ends of wooden shafts. They also headed their arrows with European points and with others that Indian craftsmen produced from pieces of iron or brass. John Underhill explained why English soldiers during the Pequot War prevented several Dutch traders from providing the Pequots with seemingly harmless articles: "Ourselves knowing the custom of war, that it was not the practice, in a case of this nature, to suffer others to go and trade with them our enemies, with such commodities as might be prejudicial unto us, and advantageous to them, as kettles, or the like, which make them arrow-heads, we gave command to them [the Dutchmen] not to stir . . . ."[37]

By trading furs, corn, wild game, wampum and even their lands, Indians acquired the European products they craved, including enough military equipment to support frequent and increasingly bloody intertribal wars. Daniel Gookin was struck by the way European weapons had replaced traditional artifacts in the Indian military system before King Philip's War:

> Their weapons heretofore were bows and arrows, clubs, and toma-hawks, made of wood like a pole axe, with a sharpened stone fastened therein; . . . but of latter years since the English, Dutch, and French have trafficked with them, they generally disuse their former weapons, and instead thereof have guns, pistols, swords, rapier blades, fastened unto a staff of the length of a half pike, hatchets, and axes.[38]

After the coming of the white man, Indian tribes had more reasons to fight each other and better equipment for their military operations. European diseases produced radical and sudden changes in the relative military strengths of various New England tribes, creating attractive opportunities for aggression and domination.[39] The fur trade caused commercial rivalries and helped drive the Iroquois into imperialistic actions which affected New England tribes. Economic greed, primarily a result of the strong demand for trade goods, was an unsettling influence in tribal relations.[40] Although the colonists tried to reduce the hostilities between tribes, white interference probably did more to cause warfare than to settle tribal differences. When colonial authorities supported one tribe against another in a dispute, they often supplied the favored tribe with enough confidence, and sometimes with enough weapons, to attempt a solution by force. The other tribe, inflamed by English intrusion and chastisement, would be very unlikely to forgive either their Indian enemies or the English. Also, as the English settlements multiplied and the Indians lost more and more of their lands, some tribal areas became too small and game too scarce. A tribe needing more hunting territory might seize land claimed by neighboring Indians and cause a war or a series of wars.[41]

While they resented the growing presence of the English settlements and the arrogant attitudes of colonial authorities, the New England Indians welcomed products of the white man's technology.

Fig. 23. Portrait of a southern New England Indian leader, usually identified as Ninigret II, principal sachem of the combined Niantics and Narragansetts. Done in oil on canvas by an unknown artist in the seventeenth or early eighteenth century.

Weapons which they recognized as clearly superior to their own, such as the flintlock musket and the steel-edged hatchet, were the most eagerly sought items in the trade economy that soon developed. As one tribe acquired European weaponry, the others had to become similarly armed to avoid military defeat or at least a serious loss of status. The arms race which ensued brought conflict between tribes and threatened the security of the English colonies as well.

CHAPTER III

The Arming
of the Indians

The Indians of southern New England did not begin to use firearms until the 1620s, but by the beginning of King Philip's War in 1675, most warriors had flintlock muskets or carbines. They acquired these weapons through trade which was often illegal. Europeans had strong economic incentives for trading with the Native Americans, and firearms were the artifacts that Indians most desired. Despite attempts by authorities in all the New England colonies to halt or restrict the sale of arms and ammunition to Indians, traders of various nationalities supplied the tribes of the region with vast amounts of the latest European munitions.

The arming of the Indians was primarily the result of the great demand for beaver in Europe. Competition among fur traders from all colonies of the northeast made it inevitable that some individuals would offer Indians the weapons they sought in exchange for their beaver pelts. During the second quarter of the seventeenth century, French, Dutch, and English colonists traded firearms and ammunition through illegal channels. Using widespread networks of intertribal trade, Indians then distributed these munitions over a great area.[1]

The acceptance of wampum as a medium of exchange, as described in Chapter II, facilitated the trade in weapons and ammunition. These beads, which could be traded for furs or any other commodity, were particularly important in the arming of the southern New England tribes, for the number of beaver in their region was relatively small. We have already seen that the production of wampum allowed local tribes to become heavily involved in trade with the Europeans. William Bradford blamed wampum and the greed of white traders for the early traffic in munitions to the Indians near Plymouth:

[Wampum] . . . fills them with pieces, powder and shot, which no laws can restrain, by reason of the baseness of sundry unworthy persons, both English, Dutch and French, which may turn to the ruin of many. Hitherto the Indians of these parts had no pieces nor other arms but their bows and arrows, nor of many years after; neither durst they scarce handle a gun, so much were they afraid of them. And the very sight of one (though out of kilter) was a terror unto them. But those Indians to the east parts, which had commerce with the French, got pieces of them, and they in the end made common trade of it. And in time our English fishermen, led with the like covetousness, followed their example for their own gain.[2]

Most early New England colonists believed that a well-armed Indian population was a real threat to the security of the settlements. The complaints of Sir Fernando Gorges and others resulted in a royal proclamation by James I on November 6, 1622, "prohibiting interloping and disorderly trading to New England in America." The proclamation expressed outrage over men who "did not forbear to barter away to the savages, swords, pikes, muskets, fowling-pieces, match, powder, shot and other warlike weapons, and teach them the use thereof." The king ordered no one to trade with the Indians without license from the Council for New England and gave that body the authority to take action against offenders. Neither this proclamation nor another like it by Charles I in 1630 could stop illegal arms trading, which was just beginning in southern New England.[3]

In 1627, the Pilgrims at Plymouth accused Thomas Morton of numerous transgressions with the Indians, including the sale of firearms and the hiring of Indian hunters. Bradford, who hated Morton and his eccentric lifestyle, claimed that the man taught Indians "how to use them [firearms], to charge and discharge, and what proportions of powder to give the piece, according to the size or bigness of the same; and what shot to use for fowl and what for deer." It is clear that Bradford's account was one-sided in its treatment of Morton and that it magnified the problem of armed Indians near Plymouth during the first decade of settlement. The free-spirited Morton may have traded a few firearms, but neither he nor others whom Bradford accused of offering guns for furs made much of an impact on the weaponry of the local tribes in this early period. Morton was sent back to England twice, once by Plymouth and once by Massachusetts Bay Colony, but he was released both times in the mother country. Bradford said almost twenty years later that the "mischief ' which Morton began had become "common, not withstanding any laws to the contrary."[4]

As leaders planned the development of the Massachusetts Bay Colony, they were aware of the activities of illegal traders along the New England coast. In April 1629, Governor Craddock in England wrote to Captain John Endicott, who was in charge of the settlement at Cape Ann, advising him to capture all who tried "to arm the Indians against us, or teach them the use of arms." Craddock wanted such offenders sent to England for "severe punishment." Once the colony was fully established, Puritan authorities administered their own justice. In 1632, the general court handed down a cruel sentence to a convicted trader in firearms: "It is ordered, that Richard Hopkins shall be severely whipped, and branded with a hot iron on one of his cheeks, for selling pieces and powder and shot to the Indians."[5]

Although the general court of Massachusetts Bay considered exacting the death penalty for further offenses like those of Richard Hopkins, a fine or a whipping was the usual punishment in later years. A law passed by the court in 1637 forbade not only the sale

Fig. 24. An Indian hunting deer with a firearm.

of all military weapons, ammunition, and armor to Indians, but also the repair of "any gun, small or great, belonging to any Indian." The penalty for each offense was a ten pound fine, but the court of assistants could increase the fine or "impose corporal punishment," if the offender could not pay. The courts could be lenient, as in the case of Robert Gowen, who "sold a gun to the Indians," but managed to get half of his ten pound fine remitted in 1650. However, several men, like George Adams in 1653, "having nothing to satisfy the law" were whipped for their crimes. Court records imply that Massachusetts Bay authorities apprehended few men for illegal trade of munitions to the Indians.[6]

A revision of the New Haven Colony laws in 1645 resulted in the following precise restrictions, which were much easier to enact than they were to enforce:

> It is ordered that whosoever shall furnish any Indian directly or indirectly either with any gun, great or small, by what name soever called or with any sword, dagger, rapier or the blade of any of them, arrow head or other weapon or instrument for war, or with any powder or shot of what name or size soever, or shall mend any gun for an Indian, without express order from the governor or commissioners for the colony in writing, shall pay either £5 fine, or tent for one, according to the nature and importance of the offence as the court shall judge meet.[7]

During the first half of the seventeenth century, Connecticut and Plymouth Colonies also passed comprehensive laws against trading firearms and ammunition to Indians or repairing their weapons. The first court action in the *Colonial Records of Connecticut* prohibited the sale of munitions to Indians "under such heavy penalty" as the court should "think meet." Connecticut authorities caught a few men for violations of the restrictions on trade, but the harshest penalties imposed were moderate fines. Although all of the colonies apparently had great difficulty apprehending men responsible for providing the Indians with munitions, the colonial courts seldom punished convicted violators severely by seventeenth-century standards.[8]

Both the seller and the buyer in an illegal sale of firearms or ammunition had good reason to keep the transaction a secret. Clandestine trading went on within the boundaries of every colony, and seamen dropped anchor at many places along the coasts of New England to make a quick profit with a cargo of munitions for the Indians. Bradford explained in verse how hard it was to stop the traders:

> Good laws have been made, this evil to restrain,
> But by men's close deceit, they are made in vain.
> The Indians are nurtured so well,
> As, by no means, you can get them to tell,
> Of whom they had their guns, or such supply,
> Or, if they do, they will feign some false lie:
> So as, if their testimony you take
> For evidence, little of it you can make.
> And of the English, so many are guilty,
> And deal under-hand, in such secrecy,
> As very rare it is some one to catch,
> Though you use all due means them for to watch.[9]

By the mid-seventeenth century, the efforts of colonial authorities to stop the arming of the Indians had failed miserably. Roger Williams, writing from Rhode Island in 1655, told the general court of Massachusetts Bay that "the barbarian all the land over" were receiving "artillery and ammunition from the Dutch, openly and horridly, and from all the English over the country (by stealth)." He thought that "no law yet extent" in either colony was "secure enough against such villainy."[10] During the following year, the Bay Colony even hired detectives in a desperate attempt to bring traders of munitions to justice. Joseph Wheeler and Thomas Hinksman were "employed by the country to find out those that sold powder, shot, and strong liquors to the Indians."[11] How successful they were in their investigation is unclear, but the Indians continued to acquire more and more arms in all the New England colonies.

Most New England colonists felt that the French and Dutch were the major culprits in the arming of the Indians. The English were indignant over the illegal actions of their own countrymen who traded in munitions, but they accused men from foreign colonies of operating on a much larger scale. This pattern of accusations probably reflects a classic case of projection; the colonists could not bring themselves to admit that the Indians of their region might have acquired at least as many of their firearms from English as from Dutch or French sources. It was less painful to focus on the transgressions of other nationalities and on the failures of other colonial governments. The governments of New Netherland and New France seemed to be allowing open trade in firearms and ammunition despite their own laws restricting such dealings. New England colonists knew that munitions passed easily from neighboring colonies into their area, and they had no effective means to prevent this trade.[12]

French trading practices enraged many English colonists who shared William Wood's opinion that a Frenchman would "sell his eyes . . . for beaver." Bradford, as noted above, blamed the French for beginning the sale of firearms to the New England Indians.[13] The French did provide firearms for many Indians in the northeast. In most of their early trade they dealt openly with practically any Indians willing to meet their price for a weapon, but by the mid-seventeenth century, French authorities were making an effort to limit sales of munitions to only those who had converted to Roman Catholicism. Even if traders had obeyed this restriction, and many did not, it would not have hurt French trade seriously. The Jesuits made thousands of "converts," and these Indians often served as middlemen, passing on munitions to any tribe that would supply furs or wampum in return. The greatest limitation on the sale of weapons by the French was not official restrictions but the steep prices of French firearms. Monopolistic companies, taxes, and high costs of manufacturing and transportation inflated the price of French muskets. The traders of New France would have sold many more firearms if their prices had not been above those of the Dutch and English.[14]

Although French trade with the Indians was a serious problem for the English, Dutch traders operating from the Hudson River valley probably did more than the French to arm the Indians of southern New England. The Dutch posts were very close to the English colonies, and the director and council of New Netherland exercised little control over the actions of avaricious private traders within their colony. Ordinances prohibiting trade in munitions were usually not enforced, particularly the 1639 ordinance which set a death penalty for that offense. By 1649, the demands of Dutch merchants forced the legalization of some trade in munitions. During the next decade, Dutch officials in New Netherland gave up almost all efforts to regulate the sale of firearms and ammunition to the Indians.[15]

In the Pequot War of 1637, the English found that a few of their Indian opponents had firearms. Although the Dutch did not trade large numbers of muskets to New England Indians until the 1640s,

Edward Johnson assumed, probably correctly, that the Pequot firearms "were purchased from the Dutch." When the English later feared trouble from the Narragansetts in the 1640s and 1650s, they again accused the Dutch of arming Indian tribes in southern New England. Many Englishmen thought that the Dutch meant to use the Indians as allies in a war against the English colonies.[16]

In 1643, Massachusetts Bay, Plymouth, Connecticut, and New Haven Colonies formed a confederation called the United Colonies of New England. Because the major purpose of this confederation was mutual protection, the commissioners of the United Colonies gave much attention to the dangerous practice of selling munitions to the Indians. In 1644, they suggested that each colony consider prohibiting the sale of weapons and ammunition "either to the French or Dutch or to any other that do commonly trade the same with Indians." Connecticut legislators acted almost immediately on the commissioners' suggestion and forbade anyone in their colony to sell "any instrument of war" to a Dutchman or a Frenchman. The commissioners took a more drastic step in 1650, when they drew up a law prohibiting foreigners from trading anything with Indians in the New England colonies. The colonial legislatures enacted the law much to the outrage of the Dutch governor in New Netherland.[17]

The commissioners of the United Colonies corresponded frequently with officials of New Netherland in order to force them to take action against traders of munitions in their colony. In 1647, the commissioners wrote to Governor Stuyvesant congratulating him on his new position, but also complaining of "the dangerous liberty taken by many [Dutchmen] in selling guns, powder, shot & other instruments of war to the Indians not only at . . . Fort Aurania [Fort Orange, later Albany] . . . but at Long Island, within the river of Connecticut, at the Narragansetts and other places within the English jurisdictions." Admitting that the Dutch might have adequate laws already, the commissioners nevertheless warned Stuyvesant that the "temptation by an excessive gain" from the sale of munitions was "so strong . . . that without a constant care and severe execution" nothing could be done. They said that they knew the difficulties of stopping the illicit trade from "daily experience" with the problem.[18]

In his letters to the commissioners of the United Colonies and other New England officials, Stuyvesant condemned the sale of munitions, "that damnable trade with the Indians," but admitted his

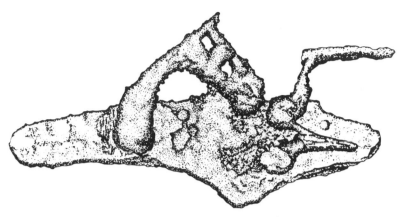

Fig. 25. Flintlock from a seventeenth-century Wampanoag burial in Rhode Island, at Burr's Hill. It appears to be Dutch.

failure to halt it entirely. Although he tried to appear energetic in his efforts to catch violators of the trading laws, neither his letters nor his actions convinced English colonial leaders of his good intentions. He first complained of the difficulty of finding guilty traders and then demanded additional evidence when the English named specific traders or smugglers.[19]

By the early 1650s, Stuyvesant was particularly sensitive to English accusations. In 1653, the New England colonists were so aroused by persistent rumors of a Dutch conspiracy with the Indians that he barely avoided war with the English. He replied to a series of English grievances and accusations, including what he called "the general complaint against trade of prohibited goods as muskets, powder and lead with the Indians." Some regulated trade of arms and ammunition was legal in New Netherland by then, and Stuyvesant knew that unlicensed traders were ignoring his restrictions with impunity. He told the commissioners of the United Colonies, "We neither will nor can allow or wholly excuse our nation therein." However, he claimed that the English were also guilty, because some of them brought "great quantities of guns and powder" into Dutch territory each year "by stealth." As an excuse for the infrequent prosecution of Dutch traders, he added: "While these things are undiscovered they can not be punished."[20]

By the 1660s, New England colonists began to doubt that the complete arming of the Indians could be prevented. England took

New Netherland from the Dutch in 1664, but numerous Dutch merchants, particularly at Albany, continued their lucrative trade in munitions with Indians in New York and New England. French and English traders also continued to arm the Native Americans at an increasing rate. Some pragmatic New Englanders argued that if the Indians could get weapons and ammunition illegally, the authorities should license honest merchants to sell munitions and let the colony benefit from the increased trade.[21]

Despite official hostility to the arms trade, prohibitions were not absolute in the English colonies. Legal sales, gifts, and loans of arms and ammunition to the Native Americans had been possible in special cases since the 1640s. The general court of Massachusetts Bay Colony frequently authorized men to supply munitions to particular Indians considered deserving of special favors.[22] In March 1644, the court made the following exceptions to its law against arming Indians:

> It is ordered, that Benedict Arnold may sell some powder and shot to Pumham and Sochonocho, provided that he let them not have above three pound a piece through the whole year; and it shalbe lawful for the said Benedict to imploy that Indian which he keepeth as his servant to shoot in a piece for the use of him, the said Benedict, so long as he shall continue in his family.[23]

In 1647, the commissioners of the United Colonies tried to stop the granting of "warrants and dispensations . . . either by the general court or by some magistrate of . . . Massachusetts to shop-keeps, or others to sell, lend, or furnish some Indians [with munitions] either because they are confederates, or under the government of that jurisdiction, or some other respect." The commissioners feared that the Indians were getting a large amount of English weapons and ammunition through these legal sources. They requested that the general court recall its warrants and in the future allow no arming of the Indians "upon any pretence or respect whatsoever."[24]

Just two years after the commissioners protested the situation in Massachusetts Bay, they allowed the general court to make another exception to the colony's restrictions on the trade of munitions. The court wanted to allow two Indians to have "two guns . . . with two pounds of powder, and six pound of shot, yearly." John Eliot, the famous missionary, personally vouched for the faithfulness of the two Indians in question. When the court submitted the proposition to the comissioners, they agreed to let the court grant the munitions once, but not yearly.[25]

In Plymouth Colony by mid-century, many settlers made a common practice of furnishing Indians "with guns, powder and shot to kill fowl and deer etc." They pretended to hire the Indians as servants for a month or more to justify arming them. In 1651, the colony prohibited this practice on penalty of a fine but still allowed men to furnish arms to Indians who had "been servants for divers years and [were] in a good measure civilized and approved of by the governor and assistants." The attitude of the Pilgrims had relaxed since the days of William Bradford's tirades against Thomas Morton and his Indian hunters.[26]

Several times in the early 1660s, the council of Massachusetts Bay Colony suggested that the general court allow the licensing of traders who would sell firearms and ammunition to the Indians, keep proper records on their sales, and pay taxes on each transaction. In the proposal sent to the general court on October 21, 1663, the council noted that the colony's laws were not restraining the Dutch, French, or English in neighboring colonies or even men in Massachusetts Bay from supplying the Indians with sufficient munitions. The general court disapproved of this proposal and of others like it in the same period, but the argument for legally arming the Indians was becoming very persuasive. Threats and actual attacks on local bands by the Mohawks were enough reason for the Reverend John Eliot to get a supply of ammunition for the "Praying Indians" he supervised.[27]

While Massachusetts Bay officials considered the justification for a legal trade in munitions, Plymouth Colony was preparing to act. In 1665, Plymouth became the first New England colony to repeal its law against selling powder and shot to the Indians. However, the open sale of ammunition in Plymouth lasted only until 1667, when the former prohibition was revived.[28]

When the general court of Massachusetts Bay finally decided to establish a regulated trade in munitions in 1668, the open arming of the New England Indians began on a large scale. The court allowed all licensed fur traders "to sell unto any Indian or Indians, not in hostility with . . . any of the English in New England, powder, shot, lead, guns, . . . [and] rapier or sword blades." The court also set the taxes that traders had to pay the colony for each sale. After the action of Massachusetts Bay Colony, both Plymouth and Connecticut

Colonies had to reexamine their policies on the arming of the Indians.[29]

The general courts in Plymouth and Connecticut Colonies were unwilling to allow the sale of both firearms and ammunition to the Indians, but the colonies could not let Massachusetts Bay take away much of their trade with the Indians. In 1669, both colonies legalized the sale of ammunition to Indians.[30] In Connecticut, a realistic explanation of the colony's dilemma accompanied the court order:

> This court upon consideration that our neighbor colonys of the Massachusetts and Plymouth have opened the trade of selling powder & lead to the Indians, and that the Indians do gain a full supply thereof according to their needs notwithstanding our law prohibits the same amongst the people of this colony, and they thereby are deprived of the benefit and gain that might be made thereof, do therefore order that for the future (and until the governor or assistants do order otherwise,) there shall be one man in every plantation in this colony appointed and approved by the respective towns to sell powder and lead to the Indians.[31]

Difficulties with the Wampanoags in the early 1670s made Plymouth's legislators reverse their policy on ammunition once again. In 1671, the general court stopped the sale of powder and shot in the colony. The authorities even tried to force the Wampanoags to surrender all their arms, but Philip and his warriors successfully avoided full compliance with the wishes of the colony. After three years without any further trouble, the general court again allowed sales of ammunition in Plymouth.[32]

Hundreds of New England Indians were already armed with flintlocks before the weakening of restrictions on the arms trade in the late 1660s. The flood of munitions to the Native Americans between 1668 and the beginning of King Philip's War seven years later left most warriors well-equipped for combat. Once that war had begun, it was easy for the colonists to believe that the insurgent tribes had been stockpiling arms for years in preparation. Although such long-range planning seems unlikely, there was evidence that the Wampanoags had made efforts to increase their store of munitions because of threatening actions by Plymouth Colony since 1671. In 1676, William Harris described the concern which the English had felt before the war: "The intention of the Indians is also revealed by their accumulation of powder, shot, and arrows. The English perceiving this, and inquiring about it, the Indians pretended it was a preparation against the Mohawks, but actually it was aimed at the English."[33]

Although colonial authorities in New England tried for many years to prevent sales of munitions to the Indians, they seldom attempted to stop them from owning or using firearms acquired in illegal trading. The futile effort to disarm the Wampanoags in 1671 was an unusual action. In 1642, the general court of Massachusetts Bay, fearful of an Indian attack, ordered the seizure of weapons and ammunition which English traders in Maine had sold to the Indians. Within a few days, the court relented and "ordered that all the Indians arms should be restored to them again." The Indians in this case were thus free to keep arms despite the law against their sale. Only in times of great fear over an Indian conspiracy did colonial authorities try to disarm warriors, and such drastic action was both uncommon and short-lived.[34]

As long as Indians did not disturb or threaten colonists, they had little trouble with the authorities over the use of firearms. Indians could not carry their muskets into Connecticut towns in the late 1650s, but in 1661, the general court allowed the Native Americans along the Connecticut River to come into towns with firearms "provided they are not above 10 men in company." In 1656, Plymouth authorities ordered that the Indians living near English towns cease all shooting at night and not disturb the quiet on Sunday with their loud weapons.[35]

For a few years one colony went further than simply allowing Indians to use firearms. In 1652, the general court of Massachusetts Bay Colony ordered certain Indians to train with muskets in the militia. Although colonists had Indian allies in the Pequot War, no New England colony had specifically trained Indians in the use of firearms before 1652. The threat of war with the Dutch and their Indian allies undoubtedly stimulated the action of the Bay Colony. From 1652 to 1656, all male Indians from sixteen to sixty living with or working as servants of Massachusetts colonists had to attend training days.[36]

With or without any formal training, Indians proved to be remarkably adept in the shooting of firearms, putting the weapons to effective use first in hunting and then in warfare. The military potential of this new technology was not lost on these Native Americans, who had their own ideas about how to fight in the forests. Muskets clearly made it easier to hit and possibly kill an opponent at a distance. The adoption of firearms by war parties was a powerful destabilizing influence in the dynamic equilibrium of intertribal relations. Those tribes who were first to acquire a stock of firearms sometimes gained a temporary advantage over their neighbors, but expanding trade networks were soon providing European armaments to Indians throughout southern New England. In 1645, the Narragansetts, who had perhaps thirty firearms, inflicted a serious defeat on the Mohegans, who had very few. Uncas, sachem of the Mohegans, immediately blamed his losses on the Narragansetts' muskets and accelerated the arming of his men.[37]

In the period between the Pequot War and King Philip's War, hostility and open conflict between major tribes and between lesser tribes and bands were constant problems for the colonial authorities. Firearms added to the problems by upsetting balances of power, by creating overconfidence among well-armed groups, and by raising the stakes of warfare. The Indians literally paid for their advanced military technology with their own blood; deaths in combat rose because of the improved weaponry. In 1669, Chekatabutt, the principal sachem of the Massachusetts, raised a large force from several tribes for an expedition against the Mohawks. According to Daniel Gookin, six or seven hundred men marched into the Mohawk territory and, after wasting most of their ammunition in a futile attempt to take an enemy fort, were caught in an ambush on their way home. The Mohawks, hiding in thick swamps on either side of the trail, "fired upon them, and killed and wounded many at the first firings." Most of the leaders, including Chekatabutt, lost their lives during the initial ambush or the ensuing battle. Gookin considered the losses to be very severe: "About fifty of their chief men, they confess, were slain in this fight; but I suppose more; but how many the Maquas [Mohawks] lost, is not known."[38] Such heavy losses in a single action were unheard of before the arrival of the white man and his weapons.

The Indians of southern New England on the eve of King Philip's War were well equipped to fight in the forests and had acquired experience with the muskets, carbines, and pistols they received through legal or illegal channels. Some were loyal to English colonies or at least submissive to colonial power; others hoped that their new weapons would enable them to throw off foreign authority and retake their lands. All had become dependent on the military technology of the Europeans and had adapted their tribal military systems to fit its capabilities and its requirements.

CHAPTER IV

Proficiency with Firearms: A Cultural Comparison

A man and his firearm form a weapons system in which the skill of the man is at least as important as the inherent accuracy of his weapon. Although the English colonists brought firearms with them to New England, very few of the men could shoot well. The average colonist had little or no familiarity with guns when he arrived and was slow to develop any proficiency as a marksman in the New World. Hunting practices and militia training on both sides of the Atlantic did not prepare colonists to shoot accurately and quickly at evasive targets. Indians, on the other hand, were trained from childhood in hunting and military skills that were readily adaptable to the use of firearms. The warriors of the forest quickly demonstrated superior abilities with the projectile weapons of the Europeans.

Any comparison of the shooting skills of the New England Indians with those of the English colonists requires some understanding of the experience that colonists may have had with firearms before coming to the New World. The abilities and attitudes of transplanted Englishmen were strongly influenced by what they had seen and done in their mother country. The residual effects of their English cultural heritage, reinforced by constant communication and continuing immigration, remained strong in succeeding generations. The colonists' proficiency with guns in the seventeenth century is, therefore, best examined with a transatlantic perspective.

In sixteenth- and seventeenth-century England, men rarely acquired ability with firearms through hunting experiences. Hunting was the sport of the upper classes and was forbidden to most of the common people. There were numerous restrictions on who could hunt and on what methods could be used; hunting laws were usually designed to preserve game for the privileged few. The most socially respectable forms of the sport were riding behind a pack of hounds, coursing with swift greyhounds, and hawking. In these activities, the hunter was relying on animals to kill other animals and was not testing his marksmanship. Bows, still common under Queen Elizabeth, soon went out of style as hunting weapons, and firearms did not immediately replace them.[1]

By law, a man needed a yearly income of at least one hundred pounds to take a firearm into the fields, forests, or marshes. Those who qualified were, however, likely to scorn the use of such an unsporting weapon. Flintlock fowling pieces did not become popular until the second half of the seventeenth century. In "The King's Christian Duties," King James I advised his son that "hare hunting, namely with running hounds" was the "most honourable and noblest" way to hunt. He warned him that it was a "thievish form of hunting to shoot with guns or bows."[2]

In 1611, John Winthrop, the English Puritan who later became governor of the Massachusetts Bay Colony, listed his reasons for giving up hunting with a firearm in England:

> Finding by much examination, that ordinary shooting in a gun, etc: could not stand with a good conscience in my self, as first, for that it is simply prohibited by the law of the land, upon this ground amongst others, that it spoils more of the creatures than it gets: 2 it procures offence unto many: 3 it wastes great store of time: 4 it toils a man's body overmuch: 5 it endangers a man's life, etc.: 6 it brings no profit all things considered: it hazards more of a man's estate by the penalty of it, than a man would willingly part with: 7 it brings a man of worth and godliness into some contempt: —lastly for mine own part I have been crossed in using it, for when I have gone about it not without some wounds of conscience, and have taken much pains and hazarded my health, I have gotten very little but most commonly nothing at all towards my cost and labour.[3]

Winthrop had taken up hunting with a gun long before the practice had any respectability. He had disgusted his neighbors with his illegal activity and had gotten no satisfaction from it. His lack of skill made hunting an unprofitable waste of time, and for a man of his convictions that was close to sin. It is doubtful that many other English Puritans had even tried the sport before the 1630s.[4]

Fowling with firearms lost its negative image only gradually.

Fig. 26. Hunting deer in seventeenth-century England, using hounds.

Some of the upper class began to fire small shot at flocks of birds with long-barreled fowling pieces, which they rested on supports. Until late in the seventeenth century, shooting at flying birds was unusual; the targets of choice were sitting and unaware of their peril. The objective was to kill as many as possible with a single discharge. This type of hunting, still restricted by law to the wealthy, did more to perpetuate bad shooting habits than to improve English marksmanship.[5]

The colonists who came to New England were, in general, not members of that upper level of English society which was beginning to hunt legally with firearms. Most were too upright to have done much hunting outside the law, and poaching game on another man's land would have been beneath their dignity. Actually, even the lowly

and disreputable English poachers had little experience with firearms; they relied on nets and traps of various types rather than noisy
fowling pieces, which would have revealed their presence. The
prospective colonists were most often busy farmers or tradesmen
whose shooting experience, if any, was limited to training they might
have received in the militia.[6]

Militia units had been training regularly in England since 1573,
when Queen Elizabeth ordered the organization of "trained bands."
All healthy men from sixteen to sixty had to defend their country if
called on, but only a select number had military equipment and
actually served in these bands. The original instructions to each
county ordered a muster of all able men, "and out of that total and
universal number being viewed, mustered and registered to have a
convenient and sufficient number of the most able, to be chosen and
collected, to be by the reasonable charge of the inhabitants in every
shire tried, armed, and weaponed, and so consequently taught and
trained, for to use, handle, and exercise their horses, armor, shot,
and other weapons, both on horseback and on foot."[7] In 1591, the
total number of trained and equipped men in the kingdom was
estimated at 42,000.[8] However, many of this group had sadly
deficient military skills and inadequate weapons.

Changes in the basic weaponry of Elizabethan trained bands
complicated the problem of instructing men in the use of military
arms. In some counties at the beginning of the seventeenth century,
the musket and the pike had not completely replaced the longbow
and the bill, a common medieval polearm. There were usually as
many pikemen as musketeers in a trained band, and the quality of
firearms carried by the latter varied greatly. As men bought or were
provided firearms for training, a wide range of weapon types and
sizes appeared in the ranks. Many of these arms were in poor or
unserviceable condition.[9]

Although Queen Elizabeth and her council urged rigorous training with weapons and the testing of marksmanship in shooting
matches, few men wanted to spend enough time in practice to learn
the proper operation of their firearms. Men received some instruction during several days of formal training each summer, but the
ideal of frequent coaching in smaller groups throughout the year
was rarely, if ever, achieved in the trained bands. Counties usually
provided insufficient powder for effective practice, and accuracy
soon became a minor consideration for musketeers. Instructors

concentrated on the numerous motions involved in loading and firing a musket and on the basic formations and maneuvers of infantry units. With the limited time spent in actual training, few men could master even these essential skills.[10]

The early seventeenth century saw diminishing English involvement in continental wars and a reduction of the threat of Spanish invasion. When Charles I came to the throne in 1625, twenty years of peace had not improved the readiness of the militia. He deplored the general decline in training and the poor condition of arms and armor in the bands of the counties, but he could do little to effect major changes in the situation.[11] Even renewed conflict on the continent had little impact on the trained bands in the countryside. Although the militia units gained recruits from the yeomanry, many joined to avoid the press gangs that gathered up available men for overseas expeditions. The king and his council were unpopular in many areas of England and could not enforce all their attempted reforms of the militia.[12]

Despite a potentially beneficial but short-lived program of instruction by professional sergeants, the trained bands remained

Fig. 27. European musketeer with matchlock musket, rest, lighted match, ammunition, and sword. One of a series of illustrations showing the exercise of arms.

Fig. 28. One of the recommended steps in loading a musket: using the scouring stick to ram a heavy lead ball and the powder charge. In the heat of battle, men sometimes dispensed with this step, and the charge might not be seated properly in the barrel.

Fig. 29. Blowing on the tip of the match to be sure it is lighted and capable of igniting the primer powder. Note that the other end of the match cord is also smoldering, as a back-up.

poorly led and inadequately trained. The king did spur modest improvements in equipment, but there was widespread resistance to increased training. In the 1630s, political and religious opponents of the royal government became more vocal, and many of them saw requirements to drill and practice with weapons as a form of oppression.[13] The task of learning to function as a competent musketeer was particularly demanding and time-consuming.

Military manuals of the period and veterans returning from European battlefields provided experienced officers and men in the trained bands with information on the latest formations, maneuvers, and techniques of firing. The increased emphasis on heavy volleys fired without careful aiming had great effect on the training of musketeers in local English bands. This coordinated, simultaneous firing by a line of men was very difficult to achieve in units

Fig. 30. Preparing to "present" the
musket in the direction of the enemy.

Fig. 31. Presenting (leveling) the musket
and firing on command, usually as part
of a simultaneous volley. This is not the
same as aiming and then firing when
ready.

which practiced as infrequently as did English trained bands.[14]

From the late sixteenth to the mid-seventeenth century, the
importance of musketry increased tremendously in European warfare, while the need for skilled marksmanship actually declined. The
military requirements of the period called for musketeers to load and
handle their matchlock weapons with precise movements, to level
them in the direction of an enemy formation, and to fire on
command in a volley. The critical element in this massing of
firepower was the simultaneous discharge of all the muskets in one
or more ranks.

A volley or a series of them by succeeding ranks could produce
terrible carnage in a tight formation of enemy soldiers. The laws of
probability had as much influence on the dispersal of direct hits and
ricochet wounds as any conscious aiming at individuals by musketeers. The terrible hail of lead balls was devastating because of its
concentrated and sudden impact. This type of massed firepower

made a high degree of individual accuracy less important than rapidity in firing and reloading the musket. A commander wanted his musketeers to fire as part of a well-drilled team and to place their bullets somewhere in the opposing formation, not necessarily in particular opponents.[15]

A musket propelled its bullet with enough energy to kill a man at five hundred yards, but no musketeer could hope to hit anything at such an extreme range. Although matchlock muskets fired in a volley were apparently effective against company-sized formations at up to two hundred yards, as individual weapons they were not accurate enough to hit a man consistently beyond fifty yards. In actual practice most European regiments probably closed within one hundred and fifty yards before the main units opened fire; battles were fought frequently at much shorter ranges. Few sources mention the ranges of battlefield musketry, but Richard Ward, writing in 1639, said: "When the enemies battalias be approached within 6 or 8 score [a score could mean either twenty paces or twenty yards, probably the former], or less, then the musketeers are to give fire."[16]

The accuracy of a musket was limited by both the deficiencies of the weapon and the way musketeers loaded and fired it. It usually had no sights and could not be aimed precisely. Like all smooth-bored firearms, the musket did not fire a ball very well because it imparted no stabilizing spin to the projectile. The weapon also shot badly because of the seventeenth-century practice of loading it with a ball slightly smaller than the diameter of the bore. Musketeers did this so that the ball would roll quickly onto the powder charge despite accumulated residue in the bore from other firings. The loose-fitting ball guaranteed some loss of velocity and accuracy, but if a musketeer did not add wadding to hold the ball on the powder, more serious problems arose. Any lowering of the muzzle would then cause the ball to roll away from the powder and perhaps completely out of the barrel. Lord Orrery said in 1677 that musketeers "seldom put any paper, tow, or grass, to ram the bullet in." When they aimed low, some balls fell out before they fired, causing "little execution . . . though they fired at great battalions, and those also reasonably near." Actually, any bullet improperly seated in the bore would not fire true.[17]

Many, if not most, adult Englishmen who came to New England before the "Great Migration" ended in 1642 had some military training with weapons, though not necessarily with firearms. They might have served in the trained bands of the early Stuarts, but very few had any real combat experience. Press gangs for foreign expeditions usually took men of lower social and economic standing than the prospective New England colonists, and only a very small number of professional soldiers and former volunteers in continental wars joined the exodus to New England. Undoubtedly many male colonists had successfully avoided even casual service in the trained bands, which contained only a minority of all eligible Englishmen. Of those who had served in the numerous foot companies of those bands, perhaps half had carried pikes and knew little or nothing about muskets. A few had belonged to the select horse troops which trained to fire pistols or carbines at close range.[18]

Even colonists who had served as musketeers in the English trained bands were far from expert in the use of their weapons. Former musketeers had spent most of their limited hours of training learning to fumble through forty or more motions of preparing to fire a matchlock musket, as described in Chapter II. They had done little actual shooting. When they had discharged a matchlock, it was with the use of a rest to support the weight of the weapon and hold it steady during the firing process. This awkward procedure was not appropriate for the military requirements of warfare against Indians or for the flintlock muskets that would eventually become the standard weapons of the colonists.[19]

In New England, the militia system was far more efficient than it was in the mother country, but the principal objective was still to produce men capable of performing the "postures of the musket" and of firing volleys on command. Colonial militiamen, although better trained and equipped than their English counterparts, received little instruction in marksmanship. There were infrequent shooting matches on muster days in particular towns; on the only such occasion mentioned in the Watertown records, the prize went to "John Spring . . . for laying his bullet nearest the mark" but not on it.[20]

Instructors in the militias of New England colonies taught the intricacies of firing matchlock and flintlock muskets according to accepted European practices. They relied mainly on popular military manuals like Richard Elton's *The Compleat Body of the Art Military* and William Bariffe's *Military Discipline*, both of which order a musketeer to "give fire breast high," with no mention of aiming. It was probably because of such fire in the general direction

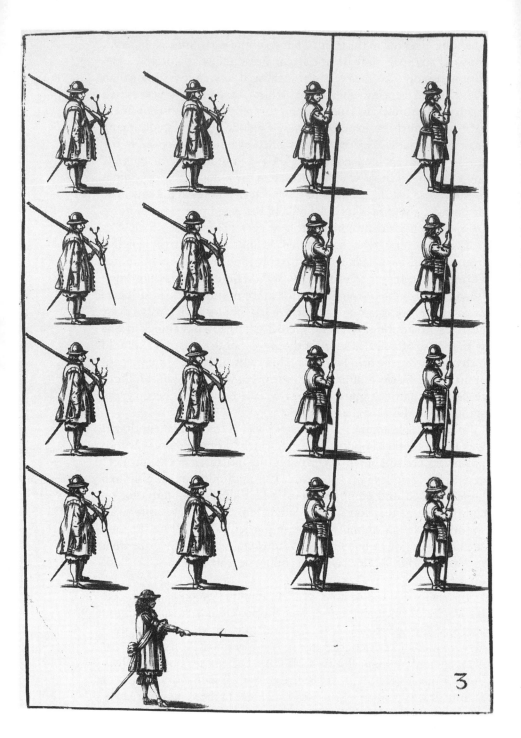

3

of the enemy that Captain Church in 1675 had to dispatch "information to the general that the best and forwardest of his army that hazarded their lives to enter the [Narragansett] fort, upon the muzzles of the enemies' guns, were shot in their backs, and killed by them that lay behind."[21]

During the Pequot War, the soldiers of Massachusetts Bay and Connecticut Colonies were confident in the merits of European volley fire. When they attacked the Pequot fort at Mystic in 1637, they began the action with a volley, evidently at the stout Indian palisade. John Underhill, a professional soldier, was amazed that "soldiers so unexpert in the use of their arms, should give so complete a volley." Everyone may have fired together for once, but the effects of the volley were superficial.[22]

Having a high regard for individual accuracy, the Indians did not share the English admiration for volleys. After the militiamen went on to burn the Pequot fort and to exact a terrible vengeance on its inhabitants, they skirmished with another group of Pequots, who were shocked by the capability of a single musket fired with care. Underhill described the incident:

Sergeant Davis, a pretty courageous soldier, spying something black upon the top of a rock, stepped forth from the body with a carbine of three foot long, and, at a venture, gave fire, supposing it to be an Indian's head, turning him over with his heels upward. The Indians observed this, and greatly admired that a man should shoot so directly. The Pequots were much daunted at the shot, and forbore approaching so near upon us.[23]

Although numerous incidents in warfare against New England Indians demonstrated the value of aiming one's musket at a single enemy, most colonial officers and musketeers did not recognize the critical importance of individual marksmanship until the last stages of King Philip's War. The Pilgrims' first skirmish with hostile Indians in 1620 should have convinced those early colonists of the need for improved accuracy. William Bradford told how an Indian shooting

Fig. 32. A formation made up of equal numbers of European pikemen and musketeers. Pikemen in the trained bands of England learned little about firing a musket. Even in the New England colonies, the pike remained an important part of militia exercises.

Fig. 33. Musketeers, with pikemen close behind them, fire a volley. At Plimoth Plantation, Plymouth, Massachusetts, reenactors demonstrate this formation on "muster days."

arrows from beside a tree "stood three shots of a musket, until one taking full aim at him . . . made the splinters of the tree fly about his ears, after which he gave an extraordinary shriek and away they went." Fifty-seven years after this occurrence, and after the horrors of King Philip's War, the authorities of Plymouth Colony finally instructed militia officers to "not only train their soldiers in their postures and motions but also at shooting at marks."[24]

The council of Connecticut Colony took steps to improve marksmanship in June 1676, when the war against Philip was turning in favor of the English. The council ordered each town to appoint someone to "call forth not only the trained soldiers but also youths under sixteen years of age, such as may be judged capable, and instruct them in handling their guns, charging and discharging at marks . . . ."[25]

Although some colonists, like the Reverend William Hubbard, still deplored any effort to "aim at single persons" in the manner of the Indians, most military men realized by 1676 that successful combat against Indians required many departures from standard European military practices.[26] Men had to pick their targets and make accurate shots when fighting in the New England forest. Years of practicing the techniques of volley fire while neglecting proper training in individual marksmanship had helped to make the colonies vulnerable to the Indian attacks of 1675.

One might assume that colonists living in or near a "wilderness" would become good marksmen because of experience in hunting, whether or not they received military training in shooting at targets. Unfortunately, our popular image of sharp-shooting frontiersmen is questionable even for the early nineteenth-century settlers of Kentucky and is far removed from the reality of the seventeenth-century colonists in New England. The latter, as we have seen, had little, if any, hunting experience when they arrived, and they were soon so busy creating villages and towns that they had few opportunities to hunt. Most became hard-working agriculturists or craftsmen, more concerned with homes, shops, and cultivated lands than with the pursuit of wild animals.[27]

Early descriptions of New England give the impression that the region was a hunter's paradise. The shores and forests abounded with game of many varieties. Vast flights of water fowl made seasonal visits, and deer, bear, moose, rabbits, squirrels, turkeys, grouse, pigeons, and other animals thrived in the area. John Pory wrote of Plymouth, "their bay in a manner is covered with all sorts of water fowl, in such sort of swarms and multitudes as is rather admirable than credible." Thomas Morton said that there were so many deer, "a hundred have been found at the spring of the year, within the compass of a mile." Although Europeans were amazed at the large quantities of wild game that were present in New England during the early years of settlement, few of the early colonists became capable hunters.[28]

William Wood admitted reluctantly that "every one's employment [did] not permit him to fowl." Recognizing the incompetence of some would-be fowlers, he wrote: "many go blurting away their powder and shot, that have no more skill to kill, or win a goose, than many in England that have rusty muskets in their houses knows what belongs to a soldier. . . ." Also, beavers and otters were "too cunning for the English" but not for the Indians, "those skillful hunters whose time [was] not so precious, whose experience-bought skill [had] made them practical and useful in that particular." Roger

Williams said that Indians armed with guns killed an "abundance of fowl," because they were "naturally excellent marksmen; and also more hardened to endure the weather, and wading, lying, and creeping on the ground, etc."[29]

Accuracy had always been important to men who hunted and fought with a bow in the forests. Most male Indians had good eyesight, fast reflexes, and years of experience with a bow. They quickly learned how to aim and fire muskets or carbines and became known for their fine marksmanship. In 1622, John Pory reported that some of the Indians on the coast of Maine had acquired firearms and could shoot "as well as any Christian." He said it was "an ordinary thing with them to hit a bird flying." Roger Williams expressed similar admiration for the accuracy of the Indians in southern New England two decades later; he called them "naturally excellent marksmen" and praised their hunting skills.[30]

Because the Indians had superior talents as hunters, the colonists let them supply much of the wild game that was eaten in the settlements. Wildfowl and deer were prized as food by both the Indians and the English. John Josselyn said that although the deer were "innumerable," there were "but few slain by the English." William Wood believed that it upset the Indians more "to see an Englishman take one deer, than a thousand acres of land." Some Indians were, however, willing to put their deer-hunting abilities to use for the English and to supply venison for a price. In a poem on New England, William Bradford wrote: "For us to seek for deer it doth not boot, / Since now with guns themselves [Indians] can shoot."[31]

As we have seen earlier, Bradford was not as sanguine in the late 1620s, when he and others accused Thomas Morton of hiring Indians to hunt with firearms. The practice, usually illegal in the early days of settlement, became common as a way of acquiring both furs and meat. Bradford later claimed that Morton instructed Indians in the use of European weapons and "employed some of them to hunt and fowl for him, so as they became far more active in that employment than any of the English, by reason of their swiftness of foot and nimbleness of body, being quick-sighted and by continual exercise well-knowing the haunts of all sorts of game." In 1634, a court in Massachusetts Bay fined two very reputable citizens, William Pynchon and Thomas Mayhew, five pounds each for "employing Indians to shoot with pieces."[32]

By the 1640s, colonial authorities were no longer as rigid in their

Fig. 34. An Indian and his highly-prized possession, a musket.

opposition to paid hunters from local tribes. Indians were providing a valuable service by eliminating wolves, which killed livestock and were seen (probably unfairly) as a menace to humans. The English had great difficulty killing these predators because of their cunning and their ability to outfight even greyhounds and mastiffs. William Wood said that wolves were the "greatest inconveniency this country hath" and that "many good dogs have been spoiled by them." In 1645, the general court of New Haven Colony complained that despite generous bounties on the troublesome wolves, no Englishman took up "this service as his employment and business." The bounties paid by towns and colonies were, however, a welcome source of money or European goods for Indian hunters. Sometimes these payments were even in the form of powder and shot.[33]

Wolves and deer were wary animals which clearly presented too great a challenge for average colonists. Some men did try their luck at fowl-shooting, a type of hunting that required no real accuracy in this period of New England history. A few shot large numbers of the waterfowl that frequented the New England coast from the fall to the early spring, but this activity took more time than most men were willing to spend away from their daily tasks. Isaack de Rasieres said that the geese at Plymouth in 1627 were "easy to shoot, inasmuch as they congregate together in such large flocks." Wood claimed that some men had killed "50 ducks at a shot, 40 teals at another." His figures may have been inflated, but they show that men shot at entire flocks and not at individual birds. Thomas Morton provided further evidence of this fact when he wrote: "I have had often 1000 [geese] before the mouth of my gun."[34]

Men also killed small birds by the dozens without any shooting skill. Wood reported: "Such is the simplicity of the smaller sorts of these birds, that one may drive them on a heap like so many sheep, and seeing a fit time shoot them; the living seeing the dead settle themselves on the same place again, amongst which the fowler discharges again. I myself have killed twelve score at two shoots." Colonists sometimes shot into groups of birds bothering the crops. De Rasieres described how this was done when he took part in the shooting at Plymouth: "Sometimes we take them by surprise and fire amongst them with hail shot, immediately that we have made them rise, so that sixty, seventy, and eighty fall all at once, which is very pleasant to see."[35]

The technique which Wood recommended for killing turkeys was to find them "towards an evening" and watch "where they perch." Then one could come back in the night and shoot as often as he wanted. The birds would "sit, unless they be slenderly wounded." For more sport a colonist could shoot flying pigeons in flocks of "millions and millions." The vast flights seemed to have "neither beginning nor ending," and "the shouting of people, the rattling of guns, and pelting of small shot could not drive them out of their course." Actually men found it more effective to take these birds with nets as they roosted than to shoot into their crowded flights. By the 1670s, John Josselyn thought the number of pigeons was "much diminished," because of "the English taking them with nets."[36]

Dogs sometimes made up for the colonists' meager hunting and shooting abilities. When winter snow had a hard crust, deer broke through the surface, but according to Wood, dogs could "run upon the top and overtake them, and pull them down." He admitted that a deer in the rugged forest could outrun even a greyhound during the summer. The dogs could not kill a bear, but Josselyn explained how they provided hunters with an easy shot: "Hunted with dogs they [bears] take a tree where they shoot them."[37]

Many colonists who came to New England expecting to take game with little effort were greatly disappointed. William Wood raved about the excellent hunting, but he also blamed new colonists for believing that they would have no trouble feeding their families in the New World. "For many hundreds hearing of the plenty of the country, were so much their own foes and [the] country's hindrance, as to come without provision; which made things both dear and scant. . . ." Weakening his argument on the ease of taking game in New England, Wood advised prospective colonists to "carry . . . much meat for a few mouths."[38]

Game was plentiful in New England, but the number of animals available and the difficulties of taking them varied with the seasons. Sometimes men could kill food easily, but at other times the early settlers almost starved when their crops were insufficient.[39] Edward Winslow tried to explain the paradox of hunger in a land of plenty:

> But here it may be said, if the country abound with fish and fowl in such measure as is reported, how could men undergo such measure of hardship, except through their own negligence? I answer, everything must be expected in its proper season. No man, as one saith, will go into an orchard in the winter; so he that looks for fowl there in the summer, will be deceived in his expectation. The time they continue in plenty with us, is from the beginning of October to the end of March; but these extremities befell us in May and June.[40]

Hunting became worse as the seventeenth century progressed. William Bradford said that waterfowl, the major quarry of colonial hunters, "decreased by degrees" after settlement by the English. Waterfowl also became more wary of shooters than they had been in the early years. In a letter to the Earl of Warwick in 1644, John Winthrop wrote: "Sea fowl here is store, but not so easy to be taken than at our first coming." A colonist named Pond was not satisfied with the shooting even in 1631. He told his father that New England had "good store of wild fowl" but said they were "hard to come by." Two years later the general court made wasting time illegal and urged constables to catch "unprofitable fowlers." The action may have discouraged hunters like Pond whose luck or skill left something to be desired.[41]

Inexperienced English hunters hesitated before entering the imposing forests of New England in the seventeenth century. It was one thing to look for fowl along the coast, but something else to go into the interior in pursuit of game. William Wood told of a man who, "ranging in the woods for deer, traveled so far beyond his knowledge, till he could not tell how to get out of the wood for trees." The Indians found him after his foot had become frostbitten during a cold night. The swamps were the most frightening places in the wilderness, and few colonists were willing to set foot in those "vast and hideous" places no matter how many animals took refuge there. John Winthrop thought that New England had a good supply of game, but he said that "the country is too full of coverts for hunting."[42]

Archaeological excavations at the sites of eight seventeenth-century homesteads in Plymouth Colony have yielded valuable clues about the diets and hunting practices of the colonists. Few wild animal bones were found among a total of approximately ten thousand bone fragments from the sites; well over ninety percent of the fragments were from domestic animals slaughtered for food. Bones from wild ducks appeared fairly often, but excavators found none from turkeys and only a very few from wild mammals. Although none of the sites dates from the first thirty years of the colony's existence, the preliminary results of this faunal analysis lend support to the theory that hunting in New England was an infrequent practice usually limited to the relatively easy shooting of wildfowl.[43]

English colonists freely acknowledged the superiority of the Indians as hunters and as marksmen. Although most thought that these skills were not worth a great deal of effort to acquire, there was a certain amount of barely-suppressed envy in the descriptions of Indian prowess with firearms. We can only guess the feelings of a group of militiamen at a certain training day some time before 1643. The officers had arranged a shooting match, which was a decided change from the normal training in marching, loading of muskets, and volley firing. At this date some or even most of the colonists would have been armed with cumbersome matchlock muskets. Each man took his turn in firing at the target, and Roger Williams was among the spectators who observed this interesting exercise. He reported that "when all of the English had missed the mark set up to shoot at, an Indian with his own piece (desiring leave to shoot) only hit it."[44]

Accuracy could have a mystical aspect in Native American culture. A piece of Mohegan folklore collected in the early twentieth century describes an incident during the Narragansetts' siege of the Mohegans' Fort Shantok. The specific references to the combatant tribes and the location connect this tale to a recorded attack in 1645. A Narragansett had climbed a tree from which he could look over the palisade of the fort and direct the fire of his fellow warriors at the defenders. According to oral tradition, the finest Mohegan marksmen were unable to hit this exposed figure, who was protected by some form of supernatural power. "At length, a Mohegan, who possessed power equal to that of the Narragansett appeared and ordered the others to desist. Taking a bullet from his pouch he swallowed it. Straightway it came out of his navel. Again he did it with the same result. Now he loaded his rifle [actually a smooth-bore musket] with the charmed ball and taking aim, fired at the man in the tree. The Narragansett dropped out of the branches, dead."[45]

The Indians' choice of firearms and ammunition had a significant effect on their ability to hit what they were aiming at, particularly if their target was as elusive as a deer or an alert warrior from another tribe. In trade they demanded flintlocks, which allowed them to "snap shoot" at a moving target. They came to prefer firearms with relatively short barrels, weapons which a man could point rapidly and which were easy to use in thick vegetation. Some even carried pistols for combat at close quarters. However, pistol balls proved to be more effective as ammunition for muskets than for handguns.[46]

Because Indians hunted with multiple shot in their muskets and had no military tradition of using large musketballs in warfare, they probably fired pistol balls or "swan shot" more often in combat than the English did. Their hunting experience would have taught them that they could kill large animals such as deer or bear with balls smaller than a standard musketball, which weighed one-twelfth of a pound. A musket could easily fire up to ten or twelve shot of sufficient size to produce a mortal wound, and the spreading out of these lead projectiles in flight greatly improved the probability of hitting a man within a range of fifty yards. A well-aimed shot at a closer distance would have a devastating effect because of the simultaneous impact of more than one ball.[47]

The colonists, like the Indians, soon recognized that the heavy shot and pistol balls used in some types of hunting had military value in both matchlock and flintlock muskets. Even if full-sized bullets were the recommended ammunition for musketeers on European battlefields, there was some deviation from this rule among certain New England militiamen. At a Watertown muster day in 1632, a member of the militia accidentally demonstrated the effectiveness of a firearm loaded with multiple shot. John Winthrop reported that the man, "having a musket which had been long charged with pistol bullets, not knowing of it, gave fire, and shot three men."[48]

William Hubbard said that one group of colonial soldiers in 1637 used "pieces laden with ten or twelve pistol-bullets at a time" to massacre some unfortunate Pequots surrounded in a swamp and "sitting close together."[49] The gruesome objective of the multiple balls in this case was not to improve the chances of hitting a single person, but to kill a number at one firing. In 1639, the general court of New Haven Colony ordered that all men bearing arms must have "20 bullets fitted to their musket, or 4 pound of pistol shot or swan shot at the least." Connecticut Colony also gave men a choice between "serviceable bullets or shot," and the supplies for a Massachusetts expedition at the beginning of King Philip's War included both "bullet and shot" in amounts sufficient for use with six barrels of powder.[50]

By the time of that 1675 expedition, the New England colonists had made numerous adaptations in their militia system, but they were still restricted by allegiance to cultural traditions and to standard military practices of their age. They did not acknowledge the obvious advantage of multiple shot over full-size musket balls in the close-range encounters of forest warfare. They had finally given up the matchlocks still in use on European battlefields and had armed themselves with the flintlocks that Indians had preferred for half a century. Yet they still did not realize the crucial significance of their flintlock muskets' potential for accurate fire at either stationary or moving opponents. Neither their militia drills nor their minimal hunting experience gave them any real training in marksmanship. It would take months of military defeat before the colonists would admit that the Indians' way of employing muskets in warfare was clearly better than their own.

An Indian in combat chose an opponent, aimed specifically at that individual, and used acquired abilities in marksmanship to kill or wound him. He often chose small balls for his flintlock and was, for that reason alone, more likely to score a hit than any musketeer firing only the standard single ball. An Indian was also more likely to be well-practiced in the aimed firing of his weapon than an English colonist. He had probably spent years snap-shooting at moving animals, testing the rapid ignition system of his flintlock and learning how to lead a moving target. He could fire his weapon from many different positions; he could even shoot with it steadied against a tree or rock that shielded most of his body from enemy bullets. He understood that he was both the potential target and the skilled operator of a deadly machine. The colonist had brought the firearm to the New World; in King Philip's War, the Indian would demonstrate how to use this machine.

CHAPTER V

Technology, Tactics, and Total Warfare

The Indian military system in southern New England underwent dramatic changes during the seventeenth century as Europeans introduced Indians to new weapons, tools, technological skills, and military philosophies. Indians selectively adopted artifacts, craft practices, and ideas to fit the perceived needs of a culture threatened by both colonial expansion and tribal rivalries. They gained confidence from their increasing understanding of European technology and lost whatever awe they may have initially felt at the sight of such European products as firearms and axes. Even in the early 1630s, William Wood of Massachusetts Bay noticed that "fresh supplies of new and strange products hath lessn'd their admiration, and quickened their invention, and desire of producing such things as they see, wherein they express no small ingenuity, and dexterity of wit . . . ."[1]

The prospect of Indians extensively adopting English material culture and practicing English crafts aroused mixed feelings among the Puritans. Although few colonists rejected the idea of teaching the Indians the Puritan faith and way of life, many recognized the danger of a well-armed and technologically skilled Indian population. William Bradford feared that improved technology would allow the local tribes to maintain and repair weapons acquired through illegal channels. Colonial authorities allowed craftsmen to instruct trusted Indians in useful trades, but a royal proclamation in 1630 forbade anyone to teach an Indian "to make or amend" firearms "or anything belonging to them."[2]

Southern New England Indians had many opportunities to learn English craft techniques. The language barrier, which frustrated the first efforts to convert the Indians, had little effect on the transfer of technology between cultures. Craft techniques can spread easily across cultural boundaries without verbal communication. Indians already versed in traditional methods of manufacturing aboriginal artifacts could learn a great deal just by observing English products and experimenting with tools and materials acquired in trade. Since they moved freely through colonial settlements, Indians could also observe English craftsmen at work and perhaps receive some instruction. Thomas Shepard said that the Indians at Nonantum in 1647 were "very dextrous at any thing they see done once." In the previous decade, Wood had noted that Indians could "soon learn any mechanical trades, having quick wits, understanding apprehensions, strong memories, with nimble inventions, and a quick hand in using of the axe or hatchet."[3]

Many Indians acquired technological information and experience while serving as part-time laborers for English tradesmen and farmers. Because labor was scarce in New England, colonists were often glad to hire Indians not only for menial chores, but also for skilled work. Members of the Narragansett tribe, despite their frequent difficulties with colonial authorities, performed well as paid workers. The tribe had a reputation for fine craftsmanship and diligence. Daniel Gookin, superintendent of Indians in the Bay Colony, called the Narragansetts "an active, laborious, and industrious people." In the early 1670s, he reported that more were "employed, especially in making stone fences, and many other hard labors, than any other Indian people or neighbors."[4]

In the 1660s, a major intertribal war drove Indian refugees into English settlements, where they had to work for their food and clothing. The dreaded Mohawks from New York invaded Indian lands in the Bay Colony and forced many of the Massachusetts and other nearby Indians to flee from their villages, fields and hunting territories. Fearful of leaving the protection of the English, the Indians became self-supporting workers in the English communities. Gookin explained that ' necessity forced them to labor with the English." Some remained for many years and must have learned valuable craft techniques as they worked beside the colonists.[5]

Indian youths served as apprentices to English craftsmen in some colonial towns. In 1660, the commissioners of the United Colonies encouraged Indian parents "to put [out] their children [as] apprentices for [a] convenient time proportional to their age to any Godly English." In return for a youth's services his parents received "yearly . . . one coat . . . besides meat, drink and clothing convenient for their children while they continue with their masters." The colonial authorities hoped the apprentices would become not only competent tradesmen, but also trusted Christian Indians.[6]

Puritan missionaries, believing that a Christian should lead a useful and industrious life, made technological training an important part of the process of "civilizing" their Indian converts. They provided both instruction and tools for Christian Indians and strongly encouraged them to learn a trade.[7] For some Indians the promise of such an introduction to English technology may have been as great an incentive for conversion as any attractions of the English faith.

In 1649, benefactors in England formed "the Society for the Propagation of the Gospel in New England," a charitable organization which collected funds to support the work of missionaries such as John Eliot, Daniel Gookin, Richard Bourne, and Thomas Mayhew. During the third quarter of the seventeenth century, several thousand southern New England Indians accepted the religion of the Puritans and tried to imitate much of their way of life. With funds from the English society and from local donors, the commissioners of the United Colonies gave missionaries books, tools, and other goods needed by the Christian Indians. Daniel Gookin reported that the commissioners provided "sundry tools and instruments for the Indians to work with in several callings." In 1653, the supply of tools granted to John Eliot for one of his 'praying towns" included felling axes, broad axes, crosscut saws, whip saws, chisels, drill braces, hoes, and grinding stones.[8]

John Eliot formed fourteen praying towns for the Christian Indians in the Bay Colony. In the early 1650s, Indians in the first praying town at Natick built a sturdy wooden bridge 80 feet long, a palisaded fort, several frame dwelling houses and a two-story meetinghouse. A carpenter assisted them for two days during the raising of the meetinghouse, and Eliot provided instructions, equipment, and encouragement for most of their other projects, including their preparations for military defense of the town. With the help of missionaries, Christian Indians learned trades, and some established commercial enterprises in their towns. They operated a sawmill at Natick, produced shingles and clapboards at Punkapoag, and spun thread in almost all the communities. One Indian even mastered enough of the printer's art to help John Eliot produce his Indian Bible.[9]

Fig. 35. John Eliot preaching to the Indians.

The most powerful southern New England tribes rejected the religion of the Puritans but still benefited indirectly from the technological education of Christian Indians. Converts living in praying towns shared some of their newly acquired craft techniques with the far more numerous non-Christian Indians around them. During King Philip's War, many Massachusetts colonists so distrusted and maltreated the Christian Indians that some of them fled to the tribes fighting the English.[10] Taking their technological skills with them, they probably found proper appreciation for their abilities among the followers of King Philip, the great Wampanoag sachem.

As artisans in the southern New England tribes learned English craft skills, many applied their new tools and techniques to the tasks of producing and maintaining military equipment. During the first half of the seventeenth century, Indian craftsmen progressed from hammering arrowheads out of pieces of brass kettles to casting bullets and making basic repairs on muskets.[11] This internal technological support for the Indian military system, combined with the growing influx of European weapons, made the New England tribes better armed as each year passed. The trend was ominous to many colonists who feared their Indian neighbors' technical advances.

In the 1640s, William Bradford was already expressing great apprehension over the advanced state of tribal technology. The

Fig. 36. Indians building a meetinghouse at Natick.

THE SKULKING WAY OF WAR

Indians had "moulds to make shot of all sorts, as musket bullets, pistol bullets, swan and goose shot, and of smaller sorts." They even owned "screw-plates to make screwpins themselves when they want them." He reported that the Indians were skillful in the use of many English tools and could "mend and new stock their pieces . . . as well in most things as an Englishman." Bradford was well aware of the military value of such technological capabilities.[12]

The Indians learned the art of casting in lead, pewter, and brass soon after the establishment of colonies in New England. They cast tobacco pipes, buttons, bullets, small shot and perhaps some of the fixtures for muskets. Because bar lead was cheaper and more available to the Indians than cast shot, they used metal or stone molds to make their own ammunition. Archaeologists have found stone molds for casting single bullets and rows of smaller shot at New England sites.[13]

Archaeological evidence also demonstrates that Indians made and resharpened gunflints for their firearms. Mohegans at Fort Shantok worked a great deal on gunflints, leaving many resharpening chips on the site. The New England Indians' long experience with stone-working would have made it easy for them to fashion gunflints and to keep their striking edges ready for action. Traditional craft skills used to produce projectile points and cutting tools from lithic materials were simply reapplied to meet the need for gunflints.[14]

Repairing firearms was a much more difficult process than casting bullets or making gunflints. Although Indians soon learned to restock muskets and substitute some new parts for broken ones, whenever possible they took badly damaged firearms to experienced English blacksmiths. However, in the 1640s, colonial authorities forbade the mending of the Indians' weapons, and tribes had to depend almost entirely on the skills of their own craftsmen. The great demand for repairs on muskets caused some Indians to become capable blacksmiths specializing in work on firearms.[15]

Knowing that some tribes had the capability to mend their own firearms, colonial authorities were often as concerned about the sale of gun parts to the Indians as they were about the trading of complete weapons. Indians may have received large numbers of muskets in a disassembled form, because parts were easier to smuggle into a colony than large firearms. In 1653, some Dutchmen concealed gun barrels and locks inside liquor casks in order to bring the articles into New Haven. The colony soon included "stocks or locks for guns"

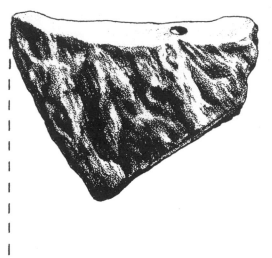

Fig. 37. Half of an Indian mold for casting six lead shot. The piece is made of stone and was found at Manchaug Pond, in Sutton, MA.

Fig. 38. Half of a shot mold found at the Burr's Hill burial ground of the Wampanoags in Warren, RI. Probably acquired in trade, it is metal and is 18 centimeters long. An Indian could cast 28 lead shot at a time using this mold.

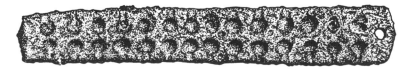

Fig. 39. Both halves of a hinged shot mold, also metal and found at Burr's Hill. One depression would cast a larger shot than the other six.

among the growing list of items that no one was to furnish to the Indians.[16]

In 1656, Joseph Jones of Massachusetts Bay violated his colony's law by giving an Indian a musket with a defective barrel and broken stock in exchange for a new barrel.[17] The Indian apparently wanted only the serviceable firing lock of the weapon. His act of trading for an obviously damaged musket was a strong indication that he or other men in his tribe had the capability to salvage useful parts for assembling whole muskets or repairing damaged ones.

Serious disputes with Plymouth Colony in 1671 caused the Wampanoags to prepare for possible military action. Hugh Cole visited the tribal headquarters at Mount Hope in that year and noted the presence of Narragansett craftsmen who were repairing the Wampanoags' firearms. These artisans were apparently experienced in working with muskets and other metal weapons. In their Rhode Island home, the Narragansetts had their own forge, and at least one member of the tribe was a skilled blacksmith.[18]

Recent excavations at a seventeenth-century Narragansett cemetery in Rhode Island have revealed an Indian blacksmith kit. This set of tools, including a hammer and some chisel-like wedges, was apparently buried with a man who had skill in its use. Narragansetts believed that souls in their afterlife would have need for many of the same artifacts they possessed when alive. Iroquois (Seneca) sites in New York from the same general period have also yielded tools for working on firearms. In addition to normal blacksmithing tools and a great many flintlock parts, Iroquois graves contained a hand vise (probably for holding sears and tumblers in filing operations), a three-cornered file, and a whetstone. One mass of 426 flintlock parts found in a single Iroquois grave must have belonged to a very capable and busy repairman.[19]

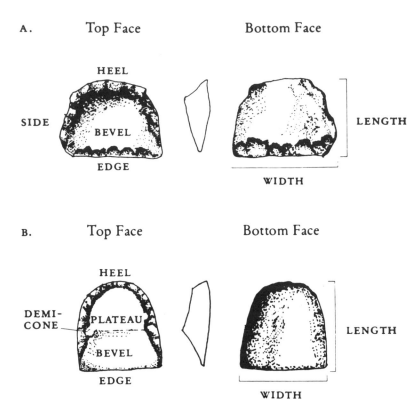

A.   Top Face   Bottom Face

HEEL

SIDE   BEVEL   LENGTH

EDGE

WIDTH

B.   Top Face   Bottom Face

HEEL

DEMI-CONE   PLATEAU   LENGTH

BEVEL

EDGE

WIDTH

Fig. 40. Two common types of European gunflints: A) Spall, and B) Blade. Indians also made flints from local stone, usually with a technique known as bifacial flaking.

One skill which the Indians lacked, and wanted desperately, was the making of gunpowder. When the Pequots, who were at war with the New England colonies, captured two English girls near Hartford in 1637, they thought they might learn the process from the girls. Edward Winslow said that the Pequots asked them "whether they could make gunpowder. Which when they understood they could not do, their prize seemed nothing so precious a pearl as before." Dutch traders freed the girls, who soon informed the colonial forces that the Pequots were short of powder for their few muskets. Despite a growing number of illegal traders in ammunition, temporary powder shortages continued to plague the Indians in New England. [20]

The English colonists did not build a successful gunpowder mill in America until 1675, when one went into operation on the

Fig. 41. Indians repairing firearms.

Neponset River in Massachusetts Bay Colony. The failure of earlier colonial efforts had not eased William Bradford's lingering fear that the Indians would somehow master the process and make their own gunpowder. Although the new mill was forced to import most of its saltpeter and brimstone during King Philip's War, William Harris of Rhode Island expected the hostile tribes in that conflict to establish their own manufacture very quickly.[21] The Indians really had no chance of doing so, but the apprehension of English observers is a good measure of their grudging respect for Indian technological potential.

Only a part of Indian military technology was devoted to weaponry; Indians also continued the aboriginal practice of building forts. The basic design and construction of most Indian fortifications remained unchanged after the arrival of the English, but Indians could build log palisades better and faster with European tools. In some cases colonists assisted Indian allies in the construction of their forts. The English may even have suggested minor modifications in this defensive technology.[22]

In the early 1640s, members of the Connecticut militia helped the Mohegans to build a fort on the Thames River. The palisaded defenses withstood a major siege by the Narragansetts in 1645. During the same period, Massachusetts authorities authorized the sending of ten men to Warwick, Rhode Island, to construct and defend "a strong house of pallizado" for Pomham, an independent Narragansett sachem temporarily allied with the Bay Colony. Christian Indians may also have received advice when they erected forts in most of their praying towns. The best English military manuals of the period contained detailed instructions for using logs, timbers and earth in fortifications, and the colonists made many defensive works of their own using these materials.[23]

The Narragansetts built the largest and most impressive Indian fort in New England. Supportive of the Wampanoags from the early phases of King Philip's War, this tribe began the construction of its huge fort in the Great Swamp of Rhode Island in 1675. The design included certain features which strongly suggest the influence of European engineering practices. William Hubbard mentioned "a kind of block house" at one corner of the wooden stockade and a "flanker" at another point. The palisade of the fort was still incomplete when a force of one thousand colonists attacked on the night of December 19, 1675, but these formidable defenses caused the English to suffer heavy casualties in the assault. At least seventy colonial soldiers were mortally wounded while taking and burning the fort. Fire, as in the 1637 destruction of a Pequot fort, made the Narragansetts' losses much higher.[24]

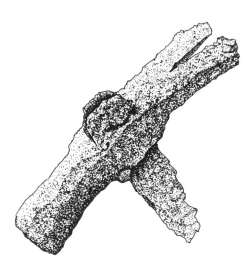

Fig. 42. A claw hammer from the Burr's Hill burial ground. Indian blacksmiths used such tools to repair broken muskets.

The tribe had successfully hidden their forge from the English until "the Great Swamp Fight." Nathaniel Saltonstall reported that the attacking soldiers killed "an Indian black-smith" who repaired Narragansett firearms. They also "demolished his forge, and carried away his tools."[25] The loss of the fort, hundreds of Indians, the forge, and perhaps their best blacksmith was a heavy blow to the Narragansetts, but many of their craftsmen escaped and joined other Indians working on the weapons used against the English.

Some Narragansetts might have been at another well-equipped Indian repair center on the Connecticut River when it was raided in May of the following year. Captain William Turner and a company of soldiers from western Massachusetts found more military equipment than they expected in the Indian camp. Saltonstall said that the colonists "demolished two forges they [the Indians] had to mend their arms; took away all their materials and tools . . . and threw two great pigs of lead of theirs (intended for making of bullets) into the said river."[26]

One band of Narraganetts did not flee the Rhode Island area but instead went into hiding in a stone fort west of the present town of Wickford. The Indians had constructed a secret refuge by using the natural boulders of a hilltop as part of their defenses and by adding connecting walls of carefully laid stone to complete the fortification. The impressive position is still easily recognizable. It is known as "the Queen's Fort," because Queen Quaiapen's Narragansett band probably built and occupied it (See Fig. 44).[27]

The Indian most qualified to build a fort of stone for the Narragansetts had shown his talent with masonry in English settlements before the war. Saltonstall wrote that he was "famously known by the name of Stonewall, or Stone-layer John; for that being an active ingenious fellow he had learned the mason's trade, and was of great use to the Indians in building their forts, etc." His engineering skill probably accounts for the sophisticated plans of both the wooden fort in the Great Swamp and the queen's stone fort. The layout of the latter structure includes a semicircular bastion and a sharp flanker, features which conform to seventeenth-century principles of European military engineering.[28]

English forces did not discover the Narragansetts' stone fort during the war, but they did surprise the female sachem and her followers at a temporary campsite in June 1676. Major Talcott's company from Connecticut killed most of the Indians, including

Fig. 43. The Great Swamp Fight in 1675. Indians are shown defending the palisade at the upper right.

Queen Quaiapen and Stonewall John. The Indian craftsman has been the subject of considerable folklore since his death.[29]

According to one Rhode Island legend which insults the New England Indians, Stonewall John was a renegade military engineer from England.[30] Some people, unaware of the level of technology among the southern New England Indians in the seventeenth century, have found it difficult to believe that he was really an Indian. His reputation has overshadowed the achievements of other Indian craftsmen with widely varying degrees of skill who performed essential services for their tribes. Stonewall John was no doubt an exceptional dry mason and architect of fortifications, but he was only one individual within an entire system of technological support. Narragansett fortifications, although impressive examples of military engineering, were in fact the Indians' least effective application of technology to warfare. Only the Narragansetts tried to defend a large fort against the English in King Philip's War; the result was the catastrophe of the Great Swamp.

The English action at the Narragansetts' Great Swamp fort in 1675, like the earlier assault against a Pequot fort in 1637, demonstrated the transmission of concepts of total warfare from Europe to

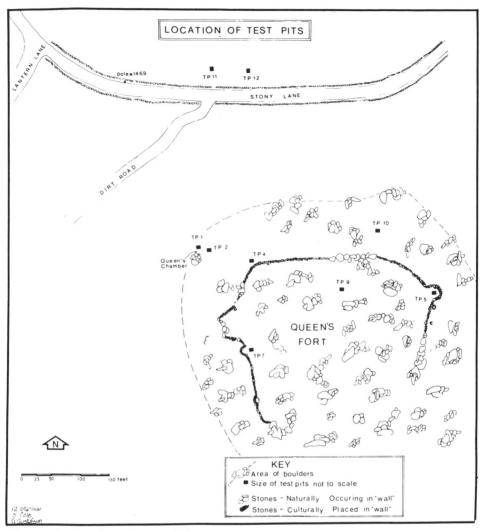

Fig. 44. *Archaeological map of the Queen's Fort.*

Within the image:

LOCATION OF TEST PITS

LANTERN LANE

pole #1469

TP 11   TP 12

STONY LANE

DIRT ROAD

TP 1
TP 2
Queen's
Chamber

TP 4

TP 10

TP 9

TP 5

QUEEN'S
FORT

TP 7

N

0   25   50   100   150 feet

H. Obadiah
J. Cole
G. Gustafson

KEY
Area of boulders
Size of test pits not to scale
Stones - Naturally   Occuring in "wall"
Stones - Culturally   Placed in "wall"

America. The religious wars which caused so much bloodshed and destruction on the Continent in the sixteenth and seventeenth centuries included attacks on both civilians and property associated with one's opponents. Religious and national differences were used to justify military actions which went far beyond the defeat of an enemy force on the battlefield. The excesses of such warfare against entire populations took root easily in the New World, where the great cultural dissimilarity between colonists and Native Americans was used to justify brutal military practices among militia units. The violence inflicted on Indians was, in fact, even more terrible and less restrained than the horrors of the Thirty Year War, where Christians fought other Christians. To many of the English who went to war against New England Indians, their opponents were more like wolves than men. The attitudes of both officers and common soldiers were strongly affected by the mortal dangers of combat against foes who seemed to possess some strange form of animal cunning, who treated prisoners cruelly, and who would not fight in expected ways. The escalation of the war against the Pequots in 1637 was accompanied by English accusations that the Indians were less than fully human. Some colonists actually suggested that the Pequots were agents of the devil.[31]

The burning of the Pequot fort may have been the first example of total warfare in New England. After a series of accusations and violent incidents and one unsuccessful punitive expedition, troops from both Connecticut and Massachusetts Bay invaded the territory of the Pequots. The militiamen landed in Rhode Island and marched overland to the Mystic River in Connecticut where the Pequots had one of their fortified villages. The total force contained about ninety colonists, sixty Mohegans, and a number of Narragansetts who had been recruited to both guide and fight. The presence of Indian allies worried some of the English participants, but the Mohegans had already proven their willingness to kill Pequots in a previous action.[32]

On May 26, 1637, the English and their Indian allies achieved a complete surprise with a dawn attack on the palisaded fortifications. The Pequots, who probably felt safe in their fort, awoke to the sound of a volley of musketry against the log walls. Immediately after opening fire, the English charged through the entrance of the fort and into the streets of the village. The colonial commander, Captain John Mason from Connecticut, led his men in setting fire to the highly flammable Pequot dwellings. Within seconds the entire village was ablaze. The English withdrew to form a ring around the doomed fort as the fire became an inferno. Mohegans and Narragansetts waited in a larger circle beyond the soldiers.[33]

These Indian allies were shocked by the horrible scene as hundreds of men, women and children perished in the blaze or were cut down as they tried to escape. In the frenzied action, the English even mistakenly wounded some of their Indian supporters. An Indian with Captain John Underhill objected strenuously to this strange and terrible form of warfare; he "cried mach it, mach it; that is, it is naught, it is naught, because it is too furious, and slays too many men." A week before this massacre, Roger Williams had sent a note to the Puritan leaders in the Bay Colony explaining the Narragansetts' request "that women and children be spared."[34]

Underhill, who commanded the small body of Massachusetts men with Mason, answered critics by insisting that the mass slaughter of both warriors and noncombatants was justified by biblical precedent:

> It may be demanded, why should you be so furious? (as some have said). Should not Christians have more mercy and compassion? But I would refer you to David's War. When a people is grown to such a height of blood, and sin against God and man, and all confederates in the action, there He hath no respect to persons, but harrows them, and saws them, and puts them to the sword, and the most terriblest death that may be. Sometimes the scripture declareth women and children must perish with their parents. . . . We had sufficient light from the word of God for our proceedings.[35]

The Narragansetts and Mohegans had no experience with this type of war. They must have been amazed and horrified by the idea of destroying an entire village. The death, in one place, of perhaps four hundred or more Pequots created a scene of incredible carnage. Underhill admitted that "great and doleful was the bloody site to the view of young soldiers that had never been in war, to see so many souls lie gasping on the ground, so thick in some places, that you could hardly pass along."[36]

Word of the merciless actions of the English militiamen in the Pequot War spread rapidly among the native population. Conservative Puritan Philip Vincent published an account of the war in London in which he claimed that the New England colonists were

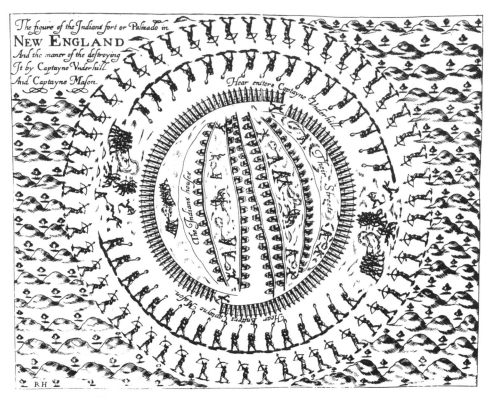

Fig. 45. A contemporary diagram of the attack on the Pequots' fort near the Mystic River in 1637. This image accompanied Captain John Underhill's account, which was published in 1638. A wooden palisade, with two entrances, surrounds a village that includes houses and streets. Flames rise from the houses, but the illustration gives little sense of the horrible conflagration caused by the colonists' use of fire as a weapon of war. Englishmen with muskets are shown shooting at the Pequots. Narragansetts and Mohegans, both allied with the English against the Pequots, are in the outermost ring of attackers. No Indians on either side have firearms.

Fig. 46. The massacre at the Pequot fort.

"assured of their peace, by killing the barbarians." He thought the war would have a positive effect as a deterrent: "For having once terrified them, by severe execution of just revenge, they shall never hear of more harm from them, except perhaps the killing of a man or two at his work . . . . Nay they shall have those brutes their servants, their slaves, either willingly or of necessity and docile enough if not obsequious."[37]

By the beginning of King Philip's War, many of the Indians in southern New England had indeed learned a lesson from the earlier demonstration of total warfare against the Pequots, but it was not the lesson that Vincent envisioned. Indians had learned that the traditional restraints which had limited deaths in aboriginal warfare were nothing more than liabilities in any serious conflict with the English colonists. Wars between Indians had become bloodier as the weapons and attitudes of the Europeans influenced the native culture. Now a great confrontation was starting, and the Indians who challenged the authority of the New England colonies were ready to fight in a new way.

The Wampanoags, Narragansetts, Nipmucs, and Pocumtucks who either joined or were swept into the war with the English and their Indian allies in 1675 followed the precedent set in the Pequot war. They waged war on all colonists, not just combatants, and they used every means at their disposal to defeat their enemies. The total warfare which the English had introduced to New England became a nightmare for frontier towns and militia bands. Although nothing that the Indians did ever approached the horror of the Pequot fort, King Philip's War showed the English how well and how fiercely Native Americans could fight.

The widespread use of fire arrows and torches against English houses was one demonstration of the Indians' new willingness to practice total warfare. Captain Thomas Wheeler, who lived through a Nipmuc attack on Brookfield, Massachusetts, in August 1675, told how the Indians wrapped special arrows with rags containing brimstone and "wild fire." Frustrated by the stubborn defense of the town's garrison house, the Nipmucs built two siege devices mounted on wheels and loaded with inflammable materials. To the relief of the trapped colonists, a rainstorm prevented the testing of this innovative equipment. Before the Indians left, however, they managed to burn every house except the defended one and one under construction.[38]

Using weapons acquired from Europeans, warriors inflicted terrible losses on military units inexperienced in forest warfare. Men trained from childhood to move silently through thickets and swamps and to wait concealed for wild game or enemies had little difficulty setting ambushes and surprising ill-prepared colonial soldiers. Before Captain Wheeler began his ordeal in Brookfield, he narrowly escaped death in the ambush of a mounted party of about twenty-five men. The troopers had ridden "in single file" down a narrow path between a "rocky hill" and a "thick swamp." Warriors lying in wait in the bushes along the path rose suddenly to send "out their shot . . . as a shower of hail." Hemmed in by the swamp before them and by a party of Indians who had let them pass, then blocked their retreat, the desperate horsemen had to climb the steep hill under fire or perish. Eight were killed and five wounded in the ambush.[39]

The so-called "Bloody Brook massacre" of September 18, 1675, was the first of several major defeats which New England colonists suffered at the hands of Native Americans. Captain Thomas Lathrop and approximately eighty men were transporting cartloads of goods near Deerfield, Massachusetts, when Indians suddenly attacked from hidden positions in a swamp. Less than ten men survived the

Fig. 47. Lancaster under attack in King Philip's War.

warriors' assault. Increase Mather claimed that "many of the soldiers [had] been so foolish and secure as to put their arms in the carts, and step aside to gather grapes, which proved dear and deadly grapes to them."[40]

The Indians accomplished two highly successful ambushes of English relief parties on a single day in April 1676. When Sudbury, Massachusetts, was the target of hostile warriors, a small relief force of about twelve reckless militiamen from Concord came to its rescue. A body of Indians surprised the group near a threatened garrison house and killed or captured all of them. A larger force under Captain Samuel Wadsworth approached the town from Marlborough, but a small number of Indian decoys lured them into the forest, where a waiting war party surrounded and badly mauled the company.[41]

Many of the militia companies organized to fight hostile tribes in 1675 and during the winter of 1676 were practically worthless in the forest because they lacked knowledgeable guides and perceptive scouts. The most capable men for these critically important duties were Indian allies. Daniel Gookin, a forceful advocate of the use of Indians by Massachusetts Bay forces, praised Indians for their ability to avoid ambushes and to track down their enemies. He ad-

Fig. 48. The Indian assault on Brookfield.

Fig. 49. Use of an innovative siege device at Brookfield. Here the Indians make fire one of their weapons against a defended garrison house.

Fig. 50. An ambush at "Bloody Brook," near Deerfield.

mired their "quick and strong sight for the discovery of anything."
Benjamin Church of Plymouth Colony supported Gookin's views on
the value of Indian scouts. Church believed that an Indian was "next
to a blood-hound to follow a track."[42] Despite the obvious need for
men with scouting skills and the ready availability of Indians
friendly to the English colonies, companies in Massachusetts Bay
and to a lesser extent in Plymouth Colony[43] did not take full advantage of the services of Native Americans during much of the war.

Military units from Connecticut Colony profited from the able
assistance of Mohegans and Pequots. These Indians were the eyes
and ears of the colony's troops as well as an alert shield between
Connecticut towns and enemy tribes. Rarely was a Connecticut
company surprised in the forest or a house in the colony burned. The
soldiers treated their Indian allies well and wondered why Massachusetts Bay could not do the same for its loyal Indians. A letter from
the Connecticut Council to that of the Bay Colony in April, 1676,
stressed the successes of units "part English and part Indian." The
council asked "why may not yourselves set out such volunteers of

Fig. 51. A scout.

both sorts and encourage, as we do, who do grant them all plunder, and give them victuals, with ammunition, and soldier's pay during time they are out. . . ?"[44]

Massachusetts Bay had a large population of Christian or praying Indians when King Philip led his Wampanoags against the English in the summer of 1675. At first Indian scouts accompanied many of the colony's units on expeditions, but not without opposition from skeptical colonists. Despite the immediate need for Indian scouts, rumors of their disloyalty spread through the towns and the military forces of the colony. Gookin believed that hostile Indians helped to start the rumors in order to deny the English the valuable assistance of the praying Indians. In any case, many colonists under the stress of war needed little excuse to condemn all Indians as untrustworthy.[45]

Ignoring the good services of Indian scouts with the army, Massachusetts Bay authorities gave in to the demands of overly suspicious individuals and sharply restricted the use of Christian Indians as allies in the war. In late August 1675, the colony disbanded its organized band of Indian scouts. Most of them had to remain in confinement with the other praying Indians of the colony. The praying towns were reduced in number and closely watched. When resentment and distrust among white colonists reached a dangerous level in the fall, officials ordered the long-suffering praying Indians interned on Deer Island in Boston Harbor. Even then a few colonists plotted to kill the people living in misery on the island. Some formerly loyal Indians refused to submit to such treatment and either went into hiding or joined the enemy tribes. Their actions added fuel to the growing hatred many colonists felt for all of the praying Indians.[46]

Fig. 52. A Connecticut unit under Major John Talcott attacking Indians in a swamp. Indian allies scouted for and fought with Connecticut's highly-successful colonial forces. Massachusetts Bay Colony was slower to acknowledge the value of Indian scouts.

A letter from Mary Pray in Providence to James Oliver in Massachusetts Bay on October 20, 1675, helped spread malicious rumors about the praying Indians. Mary Pray said that enemy warriors were boasting that "praying Indians never shot at the other Indians, but up into the tops of the trees or into the ground." She reported that when Indian scouts went into thick swamps, they told enemy warriors hiding there how they could "escape the English." To Mary Pray, many of the Christian Indians "were as bad as any other"; they took the powder which the English supplied to them and sold it to agents for King Philip.[47]

From the fall of 1675 to the spring of 1676, the forces of the Bay Colony operated with only slight help from Indian scouts. Complaints from the ranks forced some officers to dismiss scouts whom they had acquired for special expeditions. In late February, Major Thomas Savage refused to accept the command of a large force unless, in Gookin's words "he might have some of the Christian Indians upon Deer Island to go with him for guides etc." The general court consented to his demand, but the addition of six Indians to his army caused great dissent among the soldiers. Savage apparently quieted this opposition and kept his scouts. Gookin said that Savage, "being an experienced soldier, well considered the necessity of such helps in such an undertaking."[48]

Most of the officers who went into the forest without Indians to assist them either failed to find their enemies or were surprised by them in unfortunate circumstances. In October, 1675, Lieutenant Phineas Upham asked the Massachusetts Bay Council to send some men "acquainted with the woods . . . the want of which hath been a discouragement to our men." William Hubbard explained that hostile Indians continually escaped English forces in the western Bay Colony "partly by the subtilties of the enemies themselves who could easily by their scouts discern the approach of our soldiers, and by the nimbleness of their feet escape them." Hubbard suggested that "possibly if some of the English had not been too shy in making use of such of them [Indians] as were well affected to their interest, they need never have suffered so much from their enemies."[49]

While searching for an enemy party in March 1676, Massachusetts soldiers with Major Savage ignored the advice of six praying Indians whom they had reluctantly taken along "as pilots." Increase Mather noted that because of their failure to follow the direction of the scouts "the army missed their way, and was bewildered in the woods." The English not only missed a chance to trap their opponents, but also lost one man in an ambush they should have avoided.[50]

By mid-April 1676, responsible leaders in the Bay Colony began to recognize the connection between costly military blunders and the infrequent use of Indian scouts by the colony's forces. Captain Samuel Hunting received permission to raise a company of forty men from the hundreds of Indians interned on Deer Island. The warriors rushed to the aid of Sudbury on April 21, 1676, but arrived too late to prevent the slaughter of two English relief forces mentioned previously. The loyal Indians quickly became an essential part of the Bay Colony's military operations. Gookin, who had continually urged his fellow colonists to trust praying Indians, observed that after the formation of Hunting's company, "Christian Indian soldiers were constantly employed in all expeditions while the war lasted." As additional muskets arrived in the colony, the council armed more Indians and sent them to help English compa-

Fig. 53. A raid on a family in the fields. Colonists lived in great fear of surprise attacks by war parties which could move undetected for great distances through the forests and swamps. The frontier towns and any isolated farmsteads were particularly vulnerable, but few New England communities were completely secure during the first year of this costly war.

If colonists had time to barricade themselves in a fortified garrison house, their chances of surviving an Indian attack were good. Caught in the open while harvesting grain, the unfortunate farmers in this scene suffer the consequences of carelessness.

nies in the field. Their services saved many English lives and hastened the end of the war.[51]

The praying Indians were so valuable to the Massachusetts forces that by summer, English opinions of them had changed drastically. Increase Mather said that "many who had hard thoughts of them all, began to blame themselves, and to have a good opinion of those praying Indians who have been so universally decryed." Gookin thought that "impartial men" had recognized that "after our Indians went out, the balance turned of the English side." The colony did allow all the Christian Indians to leave Deer Island in May, partly because of the service they were giving in the war.[52]

Plymouth Colony also put more reliance on Indians in the spring of 1676 than during the previous fall and winter. Indian auxiliaries demonstrated their courage and loyalty with Captain Michael Pierce's ill-fated company on March 25 near Rehoboth; they suffered heavy casualties during Pierce's disastrous fight against a much larger body of Narragansetts. On subsequent expeditions the friendly Indians from Plymouth Colony conclusively demonstrated their worth and were even asked by the Bay Colony to join its own praying Indians in checking the area around the town of Medfield. By late summer, a special combined force of Indians and militiamen under the command of Captain Benjamin Church was doing great damage to hostile groups which lingered in Plymouth Colony.[53]

Until the English made good use of their Indian allies and began to adopt some Indian tactics, the warriors who opposed them were far superior in forest combat. Gookin said that the colonial soldiers were unprepared for fighting in which they "could see no enemy to shoot at, but yet felt their bullets out of the thick bushes where they lay in ambushment." The warriors sometimes camouflaged "themselves from the waist upwards with green boughs" so that "Englishmen could not readily discern them, or distinguish them from the natural bushes." Colonists "had little experience" with such warfare, "and hence were under great disadvantages." Benjamin Tompson complained in a poem that "every stump shot like a musketeer."[54]

Most disturbing to the colonists was the deadly effectiveness of Indian musketry, a result of careful aiming at individual opponents. The marksmanship and pragmatic military practices of the Indians put the leaders of militia bands in particularly perilous positions. Because in European combat the enemy usually made no effort to aim at anyone, an officer was as safe in a prominent position as were any of his soldiers in formation. The Narragansetts, who did not consider it unsportsmanlike to try to kill the leaders of their opponents, shot Captain Davenport three times during the "Great Swamp Fight." Nathaniel Saltonstall commented that "it is very possible that the Indians might think that Captain Davenport was the General, because he had a very good buff suit on at that time, and therefore might shoot at him." His distinctive dress and exposed position probably cost him his life. Six other captains were killed or died of wounds received in that action, but at least one was shot accidentally by soldiers behind him, another hazard of an officer's traditional forward position in combat.[55]

The Indian mode of warfare, actually a blend of aboriginal and European elements, proved so successful in numerous engagements that perceptive officers and government officials began to urge changes in colonial military doctrine. There was, however, considerable resistance to suggestions of adopting the Indian tactics which seemed to work so well in the forests. William Hubbard, after blaming the "Bloody Brook massacre" on the soldiers' failure to fight in a body, as expected of European musketeers, went on to criticize any imitation of the Indians' method of "skulking behind trees and taking their aim at single persons."[56]

Fig. 54. The attack on the Narragansetts' fort in the Great Swamp. The exposed position and distinctive costume of the officer at the far right would put him in great personal danger.

TECHNOLOGY, TACTICS, AND TOTAL WARFARE

Fig. 55. Governor Leverett's impressive "buff suit," which is made of very thick leather. This flexible armor could provide some protection against arrows but not against musket fire.

Fig. 56. An ambush in the forest by an Indian war party. The terrain in this scene provides cover and concealment for the ambushers and restricts the response of their opponents, who are trapped in a narrow defile.

Field commanders noted the obvious fact that their Indian opponents would not conform to European concepts of battlefield conduct. It was futile to send out a general from the Bay Colony with specific orders to stop the enemy's "skulkings whereby he picks off the English."[57] To defeat the warring tribes, the English would have to use tactics which they had long regarded with contempt and indignation. Although the process of borrowing methods from Indians was difficult for most officers and common soldiers, a few enterprising men began to depart from accepted European practices as soon as they faced well-armed and forest-wise opponents who knew the value of scouting, surprise, cover, concealment, mobility, and marksmanship. Soon a tactical and technological revolution was underway, and a new doctrine of forest warfare was evolving.

When "multitudes of Indians who possessed themselves of every rock, stump, tree or fence that was in sight" ambushed Captain Church and a party of Plymouth colonists on July 9, 1675, the Englishmen dove for any cover they could find in the barren spot where they were temporarily trapped. Church reported that a few rocks and an old wall sheltered them until they could withdraw. At the end of the same month, English forces at Nipsachuck again used natural cover in combat. Lieutenant Thomas told how his force fought Indians who were "entrenched behind trees and rocks ready for battle." The soldiers, "adopting the tactics of the enemy . . . engaged them fiercely." The Reverend John Russell of Hadley described another incident in the small tactical revolution that began in the summer of 1675. During a battle in August, Englishmen "after the Indian manner got behind trees, and watched their opportunities to make shots at them."[58]

Governor Leverett of Massachusetts Bay correctly appraised the military situation as early as September 24, 1675, when he commented that many men were "lost by not taking heed to the ambushments of the enemy nor observing their methods." He ordered the companies of the Bay Colony "to attend the enemies" method, which though it may seem a rout to ours is the best way of

Fig. 57. Captain Benjamin Church and his men adopting Indian tactics to counter an ambush near Tiverton early in the war. The colonists are taking cover behind rocks, thereby shielding themselves from musket fire while awaiting their chance to make an orderly withdrawal.

fighting the enemy in this bushy wilderness."[59] His order did not cause any immediate changes, particularly because Indian scouts had been recently removed from most of the colony's military units. However, by the end of the following summer, soldiers from all of the New England colonies were shooting at individuals, using cover when fired upon, and moving through the woods quietly and carefully. Their ability to find and attack the enemy had greatly improved as they gained experience in the forest and listened to the advice of Indian allies. Connecticut units had learned much from their Mohegan and Pequot comrades-in-arms, and Captain Church had finally gotten the mixed company of Indians and Plymouth colonists which he had wanted.

Church probably did the most to popularize the adoption of unconventional tactics by English forces. On July 24, 1676, Governor Winslow commissioned him to "raise a Company of Volunteers of about 200 men, English and Indians; the English not exceeding the number of 60." His renowned campaigns in the next two months left few doubts about the effectiveness of borrowed military methods. Church had the authority to spare captives who had not directly participated in the killing of colonists, and he accepted formerly hostile warriors into his company. He promised them that they would not be executed or "sold out of the country" (the standard treatment for captured adult male Indians) if they fought well against the remaining insurgents.[60]

This unusual Plymouth company killed or captured hundreds of Indians by fighting in the Indian manner. Church carefully studied the methods of his opponents and interrogated prisoners to gather information on enemy locations and tactics. By acting on the advice of both friendly and hostile Indians, he created a combined force which could defeat the enemy on its own ground.[61] His captives offered one suggestion which improved the movement of troops through the forest and reduced casualties in combat; they told him that "the Indians always took care in their manners and fights not to come too thick together. But the English always kept in a heap together, that it was as easy to hit them as to hit a house."[62] Crowding was very dangerous on the march, because it meant that an entire unit might be trapped in an ambush or even surrounded. If a column was properly spread out along a trail, it would take too many men to pin down the whole force, and there would be few suitable locations for such a large ambush.

Fig. 58. Captain Church, with an Indian scout in the lead.

The killing of Metacomet, or King Philip, on August 12, 1676, was a widely-publicized feat which demonstrated the value of Captain Church's Indian-like tactics. Church had learned from an informer that Metacomet was in a swamp below Mount Hope, where the war had begun. Rushing to the area with a small force, Church split his unit into a raiding party to drive the Wampanoag sachem and his few remaining followers from their camp and a blocking force to wait in ambush for any fleeing Indians. Church personally stationed his blocking force in the wooded wetlands near the Indian camp. In the semi-darkness before dawn he placed pairs of one Indian and one Englishman behind trees at spots which seemed to almost insure interception. When the raiders fired into and then rushed the camp, the elusive sachem escaped into the swamp only to run toward a pair of waiting men. It was an Indian who shot Metacomet after his English partner's musket had misfired.[63]

The ultimate defeat of the hostile tribes was inevitable long before the death of the man whom the English called King Philip. The Indians' tactical successes and their skillful use of European military technology were not enough to win a war against the far more numerous colonists, whose Indian allies, fortified garrison houses, and almost unlimited logistical support tipped the scales heavily.[64]

*Fig. 59. Paul Revere's fanciful engraving of the great Wampanoag
sachem Metacomet, also known as King Philip. This is not an
accurate representation. The Indian leader is pictured with his musket
and powderhorn standing before Mount Hope.*

Fig. 60. An Indian firing the shot that killed Metacomet.

By the fall of 1676, disease, starvation, lack of ammunition, and relentless pursuit had brought disaster to the scattered Indian survivors. The hostility of the Mohawks toward the insurgent tribes had been a boon for the colonial forces and an important factor in the English military success. There was little hope of escape to the west, where the Mohawks were waiting, and few places where the remnants of proud bands could hide in New England. Colonists had burned corn fields, destroyed food caches, and kept their enemies from traditional fishing spots. When starvation forced Indians to leave hiding places in search of food, they frequently fell prey to grim soldiers who were no longer bewildered in the forest. The English units, guided by Indian scouts, had begun to use the same ambush and raid techniques with which their opponents had terrified most of New England in the previous year.[65]

The English had also borrowed from the military materiel of the Indians during King Philip's War. Colonists learned to paddle

Fig. 61. Defense of a garrison house. Fortified dwellings provided places of refuge for the populace of communities under attack. The degree of fortification varied, but these garrison houses were better equipped for defense than the average home. Most had heavy doors, shuttered windows, and supplies of sand or water for putting out fires. The best ones had thick plank walls, flankers, or surrounding palisades. Colonists could shoot from windows and through special firing ports, or "loop holes," in the walls of some garrisons. Indians in King Philip's War were not willing to accept heavy casualties in direct assaults on well-defended houses, and prolonged sieges were usually prevented by the threat of outside relief from neighboring towns. The few Indian successes against garrisons came largely as a result of surprise attacks. The fortified houses of New England may have saved the lives of hundreds of colonists, but they could not prevent the ransacking and burning of a majority of the buildings in many towns.

Fig. 62. *The capture of the Wampanoag war leader Annawon in September, 1676. Captain Church learned of Annawon's location from prisoners and was able to surprise him and his followers. Apparently tired of being on the run at this late stage in the war, several Wampanoag parties surrendered to Church's small band without a fight. Annawon was later executed by Plymouth Colony. The rock formation shown in the drawing still exists in a swampy area of Rehoboth, MA, and is known today as "Annawon Rock."*

birchbark canoes and to travel with snowshoes and moccasins in the deep snows of the New England winters. Hatchets, or tomahawks, designed for the Indian trade became standard sidearms in some units, replacing the traditional swords. Even English food proved less suitable for forest warfare than standard Indian rations. Bitter experiences with moldy bread and heavy supplies on long expeditions prompted a militia committee in Massachusetts Bay to add the following item to a list of military provisions: "fifty bushels of Indian corn, parched and beaten to nocake." Each of these hundred soldiers was to carry his supply of this nourishing and durable meal in a small bag, as was the Indian practice. [66]

Fig. 63. Indians in hiding. Suicidal resistance, surrender, starvation, or dangerous flight were the sad choices left for the losers in King Philip's War. Colonial authorities punished many Indians severely for being part of an insurrection but allowed others to return to their former lands, now under white domination. Metacomet's own son was one of a large number of captives who were sent away to be sold into slavery.

Not every colonist was pleased that Indian allies and borrowed military practices had contributed so much to the success of the English forces, but almost everyone agreed by 1677 that warfare in the New England forests required departures from conventional European methods. John Eliot recognized the military changes which had occurred since the Pequot War:

> In our first war with the Indians, God pleased to show us the vanity of our military skill, in managing our arms, after the European mode. Now we are glad to learn the skulking way of war.[67]

Conclusion

The Indians of southern New England understood the military importance of technology long before their first meeting with a white man. Their own craft skills and their modes of warfare were well adapted to the natural environment of their region and to the needs of their culture. Wars among Indian bands or tribes were frequent but did not result in a heavy loss of life. The arrival of Europeans soon changed the culture of the Indians and raised the human costs of warfare.

The selective and skillful adoption of European weapons, tools, and craft techniques by Indians during the first three quarters of the seventeenth century demonstrated the Native Americans' impressive ability to understand and apply new technology. Many of the articles and materials which the Europeans introduced had immediate value to the Indians and were eagerly sought by them. The fur trade and the manufacture of wampum provided a steady flow of goods to the tribes in southern New England. Observation, experimentation, and occasional instruction led to the rapid acquisition of European craft skills. Indians were not, however, simply imitative; they retained many traditional forms and practices and creatively adapted new technologies to fit their own needs.

Indians quickly learned to use and repair firearms, the most valuable and deadly of the weapons introduced by the Europeans. All efforts to prevent or cut off the trade of muskets, carbines, and fowling pieces to the Indians were ineffective; illegal sales by French, Dutch, and English traders were never halted, and many of the colonial laws against this profitable trade were repealed before 1670. Indians dealing with white traders chose the type of firing mechanism that was best suited for hunting and combat in the forests: the flintlock. In their immediate preference for the flintlock over the matchlock, they showed more insight into the capabilities of firearms than did the English colonists, many of whom still used the inferior matchlock in the third quarter of the seventeenth century.

Accuracy with firearms came easier to the Native Americans than it did to the colonists. Trained from childhood to hunt with bows and to aim at individuals during wars in the forest, the Indians immediately took advantage of the flintlock's potential for accurate, aimed fire. They saw no reason for the European emphasis on massed volleys of musketry, a form of firing on command which could be devastating against troops in tight formation but which did little against the spread-out and often hidden warriors in a woodland ambush. Indian marksmanship and the killing power of the flintlock firearm were a deadly combination.

Flintlocks could be particularly effective when fired from positions with good cover and concealment. The New England environment, which included much rough terrain and dense vegetation, offered great opportunities to combatants who would take advantage of these natural features. Colonists had to learn from the Indians how to fight in the forests, how to use the trees, the rocks, and the brush in their tactics. The lessons were sometimes painful, but militia leaders in King Philip's War eventually recognized that Indian tactics were worthy of emulation.

Stealth, surprise, and high mobility were major assets in the irregular warfare at which the Indians excelled. Although no colonist could hope to move through the woods and swamps as easily or as quietly as an experienced Indian warrior, the English forces did become much better at forest warfare. The best military units relied on Indian allies to scout for and fight with them. They also borrowed Indian equipment such as canoes and snowshoes and ate the ground maize which was the Indians' combat ration. Indians played an important role on the English side in most of the colonists' tactical victories.

The English colonists finally crushed their opponents in King Philip's War, but not with the ease that had marked their earlier success in the Pequot War. The Indians who rose up against the New England colonies in 1675 had been exposed to the merciless concepts of European total warfare and had the improved technology and tactics to inflict heavy losses on the white populace. In their desperate attempt to save their culture and to take back their lands, the Indians abandoned most of the self-imposed restraints that had limited the death and destruction in their traditional patterns of warfare.

A complete defeat of the English was impossible from the beginning of the war. Their overwhelming numbers, their fortifications, and their vast network of technological and logistical support, extending from New England villages to manufacturing centers abroad, gave the English a great advantage over their Indian enemies. With some tribes supporting the colonies against the insurgents, the suicidal nature of the uprising should have been, and perhaps was, obvious to King Philip and his followers.

The Indians of southern New England did not drive the English out, but they left their mark on the white culture and, in particular, on the European military system which had been transplanted in the New World. The "skulking way of war" had shaken the confidence of the colonists and had forced them to adopt a new doctrine for forest warfare. In the later colonial wars and in the American Revolution, English colonists would further refine and develop this doctrine. The lessons learned from Indian warriors have been passed down through the history of the American military, and have had a significant impact on the conduct of all our wars.

Notes

NOTES TO CHAPTER I

1. For general information on American Indians see Harold E. Driver, *Indians of North America* (Chicago, 1961). For ethnographic material on Indians in southern New England see Catherine Marten, *The Wampanoags in the Seventeenth Century: An Ethnohistorical Survey*, Occasional Papers in Old Colony Studies, No. 2 (Plymouth, MA., 1970); Froelich Rainey, "A Compilation of Historical Data Contributing to the Ethnography of Connecticut and Southern New England Indians," *Bulletin of the Archaeological Society of Connecticut*, No. 3 (New Haven, April, 1936), pp. 1–89; Susan G. Gibson, ed., *Burr's Hill: A 17th Century Wampanoag Burial Ground in Warren, Rhode Island* (Providence, R.I., 1980); James Axtell, *The European and the Indian: Essays in the Ethnohistory of Colonial North America* (Oxford, 1981); William Cronon, *Changes in the Land: Indians, Colonists, and the Ecology of New England* (New York, 1983); and Bruce Trigger, ed., *Northeast*, vol. XV of William C. Sturtevant, ed., *Handbook of North American Indians* (Washington, D.C., 1978–1984). The articles by Bert Salwen, William S. Simmons, and Dean R. Snow in *Northeast* are particularly appropriate as background reading for this chapter. Alden Vaughan, *New England Frontier: Puritans and Indians 1620–1675* (Boston, 1965), pp. 26–63 has some good cultural information, but Vaughan has been corrected on many points by Neal Salisbury, *Manitou and Providence: Indians, Europeans, and the Making of New England, 1500–1643* (New York, 1982). See also Howard Russell, *Indian New England Before the Mayflower* (Hanover, N.H., 1980) and C. Keith Wilbur, *The New England Indians* (Chester, CT., 1978).

2. See William Wood, *New England's Prospect* (Boston, 1865), pp. 94–95; Adam J. Hirsch, "The Collision of Military Cultures in Seventeenth–Century New England," *Journal of American History*, LXXIV, no. 4 (March, 1988), pp. 1190–1194, 1200; and Roger Williams, *A Key into the Language of America* (Providence, R.I., 1936), pp. 184–189.

3. See Daniel Gookin, "Historical Collections of the Indians in New England," *MHC*, I, 1st series (Boston, 1792), pp. 147–149; Williams, *Key*, pp. 184–186, 202; John Josselyn, "An Account of Two Voyages to New England," MHC, III, 3rd series (Cambridge, MA., 1833), pp. 295, 309; John Pory, "John Pory to the Governor of Virginia (Sir Francis Wyatt)," in *Three Visitors to Early Plymouth*, ed. by Sydney V. James, Jr. (Plymouth, MA., 1963), p. 17; Samuel de Champlain, *The Works of Samuel de Champlain*, ed. by H.P. Biggar (6 vols.; Toronto, 1922–36), I, pp. 178–179, 458, II, 203–204; Thomas Lechford, *Newes From New England*, ed. by S. Hammond Trumbull (Boston, 1867), p. 118; Thomas Morton, *New English Canaan*, ed. by Charles Francis Adams, Jr. (New York, 1967), pp. 155–157; John Smith, *Travels and Works of Captain John Smith*, ed. by Edward Arber (2 vols.; Edinburgh, 1910), I, p. 206; Peter Thomas, "In the Maelstrom of Change: The Indian Trade and Cultural Process in the Middle Connecticut River Valley, 1635–1665" (Ph.D. dissertation, University of Massachusetts, 1979), pp. 30–31, 50–51, 167; Neal Salisbury, "Social Relationships on a Moving Frontier: Natives and Settlers in Southern New England, 1638–1675," *Man in the Northeast*, no. 33 (1987), pp. 92–93; and Hirsch, "The Collision of Military Cultures," p. 1190. Hirsch argues that economic factors were relatively unimportant as causes of war before extensive contact with Europeans began: Indians usually had plenty of land and were not particularly materialistic. He notes the important goal of "symbolic ascendancy, a status conveyed by small payments of tribute to the victors, rather than the dominion normally associated with European-style conquests." For discussion of how warfare could ease grief, see Daniel Richter, "War and Culture: the Iroquois Experience," *William and Mary Quarterly*, XL, no. 4 (October, 1983), pp. 529–532. Richter discusses the "deep cultural significance" of warfare among the Iroquois and the particular role of the "mourning-war" in assuaging grief. The Iroquois also used wars to replace their own dead with adopted captives, thus maintaining sufficient levels of population and replenishing the spiritual power lost with the death of individuals. This pattern of adoption does not appear to have been as prevalent among New England tribes.

4. Salisbury, *Manitou and Providence*, p. 48, points out correctly that there was, in purely anthropological terms, no tribal organization in New England. His use of the technically-correct word "band" creates confusion, however, because he uses it to define any "permanent, supra-familial organization" among the New England Indians. The word tribe has been used by many historians and anthropologists to refer to the largest groupings of New England Indians, even if the groups in question were loose associations of bands or single bands of considerable size and political independence. This study will use the word "tribe" in the latter

sense and will use the term "principal sachem" to identify the most important leader in such a "tribe." The reader should not infer that these "tribes" had a hereditary chief or that the sachem had absolute authority. See Driver, *Indians of North America*, pp. 340–347, for a discussion of tribal organization in the East. For a contemporary comment on Indian organization and leadership see Wood, *New England's Prospect*, pp. 89–90; Edward Winslow, "Good Newes from New England," in *Chronicles of the Pilgrim Fathers of the Colony of Plymouth 1602–1625*, ed. by Alexander Young (Boston, 1841), pp. 360–361; Francis Higginson, *New England's Plantation* (London, 1630), p. C3ᵛ; Champlain, *Works*, I, p. 335; Josselyn, "An Account of Two Voyages," pp. 308–309; Gookin, "Historical Collections," pp. 147–149, 154; and Williams, *Key*, pp. 140–142.

5. See Champlain, *Works*, I, pp. 417–418; Winslow, "Good Newes from New England," pp. 288, 359–360; Benjamin Church, *The History of King Philip's War*, ed. by Henry Martyn Dexter (Boston, 1865), pp. 98–100.

6. Vaughan, *New England Frontier*, p. 28 estimates the number of Indians in New England at 25,000 in 1600, basing this figure on James Mooney, "The Aboriginal Population of America North of Mexico," *Smithsonian Miscellaneous Collections*, LXXX, No. 7 (1928), pp. 3–4. Salisbury, *Manitou and Providence*, pp. 22–30 shows that this figure is far too low and provides a detailed discussion of the sources on population. Salisbury may, however, go too far when he uses estimates for precontact warrior populations given in Gookin, "Collections," pp. 147–149 and multiplies them by a "factor of 7.5 individuals per family." He gets a total of 135,000 Indians for only the region between the Saco River in Maine and the Quinnipiac River in Connecticut. In the first place, Gookin's figures were based on estimates made by old Indians and may contain some exaggeration; in the second place, the multiplication factor seems high. While Salisbury argues effectively that the plague of 1616–1617 nearly eradicated many of the coastal bands, this author cannot accept, without additional evidence, a pre-plague population in New England of well over 135,000 Indians. My best estimate is 80,000, a figure close to the 75,000 given in Russell, *Indian New England*, pp. 24–28. Francis Jennings, *The Invasion of America: Indians, Colonialism, and the Cant of Conquest* (Chapel Hill, N.C., 1975), p. 28, estimates a range of 72,000 to 90,000 for lower New England. There is no way to settle the argument over population because of the scarcity of source material and the possibility of significant errors in early estimates. I am indebted to Neal Salisbury for causing me to reconsider my earlier population estimates, which were admittedly too low.

7. See Russell, *Indian New England*, pp. 22–28; Salisbury, *Manitou*

and Providence, pp. 15, 27–30, 112; Vaughan, *New England Frontier*, pp. 51–57; Gookin, "Historical Collections," pp. 147–148; Henry F. Howe, *Prologue to New England* (New York, 1943), pp. 224–225; Christina B. Johannsen, "European Trade Goods and Wampanoag Culture in the Seventeenth Century," in Gibson, ed., *Burr's Hill*, pp. 25–26; and John W. DeForest, *History of the Indians of Connecticut* (Hamden, CT., 1964), pp. 63–64. My estimates of population are conservatively based on Gookin's figures. In general, I have multiplied his estimates of warrior populations by a factor of four. Jennings, *The Invasion of America*, p. 28, multiplies Gookin's figures by up to four. For the Narragansetts, I cut Gookin's original estimates from 5000 to 4000 warriors, because he may have included the eastern Niantics in the tribe. I have used the well-known term "Wampanoag" to avoid confusion; the name "Pokanoket" was applied to this group by early writers. For information on the Narragansetts, see William S. Simmons, "Narragansett," in Trigger, ed., *Northeast*, pp. 190–197.

8. See Russell, *Indian New England*, pp. 22–27; Salisbury, *Manitou and Providence*, pp. 21, 29–30, 83, 112, 207–208, 263fn.; Vaughan, *New England Frontier*, pp. 53–57; DeForest, *History of the Indians of Connecticut*, pp. 58–59; and Gookin, "Historical Collections," p. 147. Salisbury revises much of the previously accepted history of the Pequots and their relationship with the Mohegans.

9. See Russell, *Indian New England*, pp. 22–27; Salisbury, *Manitou and Providence*, pp. 21, 27–28, 148; Vaughan, *New England Frontier*, pp. 52–53, 56–57; and Gookin, "Historical Collections," pp. 147–149. For the role of the Mohawks in New England tribal and colonial affairs, see Neal Salisbury, "Toward the Covenant Chain: Iroquois and Southern New England Algonquians, 1637–1684," in Daniel Richter and James H. Merrell, eds., *Beyond the Covenant Chain: The Iroquois and Their Neighbors in Indian North America, 1600–1800* (Syracuse, 1987), pp. 61–73.

10. See Wood, *New England's Prospect*, p. 67; Champlain, *Works*, I, pp. 321, 327–328; G.M. Day, "The Indians as an Ecological Factor in the Northeastern Forest," *Ecology*, XXXIV (1953), p. 331; Mooney, "Aboriginal Population of America," pp. 3–4; Vaughan, *New England Frontier*, pp. 51–52; W.S. Hadlock, "Warfare Among the Northeastern Woodland Indians," *American Anthropologist*, XLIX (1947), pp. 206, 210, 214, 216; William Bradford, *Of Plymouth Plantation 1620–1647*, ed. by S.E. Morison (New York, 1967), p. 89; Salisbury, *Manitou and Providence*, pp. 20, 24, 27, 30–39; Cronon, *Changes in the Land: Indians, Colonists, and the Ecology of New England*, pp. 42–45; and Dean Snow, "Eastern Abenaki," in Trigger, *Northeast*, pp. 137–143.

11. See E.L. Butler, "Algonkian Culture and the Use of Maize in

Southern New England," *Bulletin of the Archaeological Society of Connecticut*, XXII (New Haven, 1948), pp. 23–24; Champlain, *Works*, I, pp. 410–411; Morton, *New English Canaan*, p. 160; William Harris, *A Rhode Islander Reports on King Philip's War: The Second William Harris Letter of August 1676*, ed. by Douglas Edward Leach (Providence, 1963), p. 23; Hadlock, "Warfare Among the Northeastern Woodland Indians," pp. 211–214, 219; Driver, *Indians of North America*, pp. 354–355; Salisbury, *Manitou and Providence*, pp. 30–39; Russell, *Indian New England*, pp. 30–31; and William Morrell, "Morrell's Poem of New England," *MHC*, I, 1st series (Boston, 1792), pp. 134–135.

12. See Wood, *New England's Prospect*, p. 76; Williams, *Key*, p. 11; Edward Johnson, *Johnson's Wonder-Working Providence*, ed. by J. Franklin Jameson (New York, 1910), p. 149; Mary Rowlandson, "The Captivity of Mrs. Mary Rowlandson," *Narratives of the Indian Wars 1675–1699*, ed. by Charles H. Lincoln (New York, 1959), p. 135; and Champlain, *Works*, II, pp. 85–86.

13. See James Axtell, "The Scholastic Philosophy of the Wilderness," *William and Mary Quarterly*, XXIX, No. 3, 3rd series (July, 1972), p. 347; Williams, *Key*, p. 120; Wood, *New England's Prospect*, p. 73; and Josselyn, "An Account of Two Voyages," p. 351.

14. See Williams, *Key*, pp. 106–110; Champlain, *Works*, I, pp. 337–338; Wood, *New England's Prospect*, p. 102; John Mason, "A Brief History of the Pequot War," *MHC*, VII, 2nd series (Boston, 1819), p. 149; Martin Pring, "A Voyage Set Out From the City of Bristoll," in *Sailors' Narratives of Voyages Along the New England Coast 1524–1624*, ed. by George Parker Winship (Boston, 1905), p. 58; Lion Gardiner, "Relation of the Pequot Warres," *MHC*, III, 3rd series (Cambridge, MA., 1833), pp. 147–148; and Gookin, "Historical Collections," pp. 152–153.

15. See Williams, *Key*, pp. 46–47, 72–73; and Frederic H. Douglas, "New England Indian Houses, Forts and Villages," *Denver Art Museum Indian Leaflet Series*, No. 39 (Denver, 1932), pp. 154–155.

16. See Williams, *Key*, pp. 46–47; Wood, *New England's Prospect*, pp. 94, 106; Giovanni da Verrazano, "Narrative," in *Sailors' Narratives of Voyages Along the New England Coast 1524–1624*, ed. by George Parker Winship (Boston, 1905), p. 19; Douglas, "New England Indian Houses," pp. 154–155; Josselyn, "An Account of Two Voyages," pp. 295–296; Gookin, "Historical Collections," pp. 149–150; Morton, *New English Canaan*, p. 138; Charles C. Willoughby, *Antiquities of the New England Indians* (Cambridge, MA., 1935), pp. 289–294; and Cronon, *Changes in the Land*, pp. 37–47.

17. See Willoughby, *Antiquities*, pp. 284–288; Wood, *New England's Prospect*, p. 94; Sylvester Judd, *History of Hadley* (Springfield, MA.,

1905), pp. 118–120; Champlain, *Works*, I, pp. 329–330; Peter A. Thomas, "Cultural Change on the Southern New England Frontier, 1630–1665," in William W. Fitzhugh, ed., *Cultures in Contact: The Impact of European Contacts on Native American Cultural Institutions,* A.D. *1000–1800* (Washington, 1985), pp. 136–137; and Douglas, "New England Indian Houses," p. 155.

18. Wood, *New England's Prospect,* pp. 94–95.

19. See Josselyn, "An Account of Two Voyages," pp. 309–310; Gookin, "Historical Collections," pp. 166–167; DeForest, *History of the Indians of Connecticut,* pp. 213–214; Judd, *History of Hadley,* pp. 118–119; and John K. Mahon, "Anglo-American Methods of Indian Warfare 1676–1794," *Mississippi Valley Historical Review,* XLV, No. 2 (September, 1958), p. 263.

20. See Pring, "A Voyage Set Out from the City of Bristoll," p. 56; James Rosier, "A True Relation of the Most Prosperous Voyage Made this Present Yeare 1605, by Captaine George Waymouth in the Discovery of the Land of Virginia" in *Sailors' Narratives of Voyages Along the New England Coast 1524–1624,* ed. by George Parker Winship (Boston, 1905), p. 119; Wood, *New England's Prospect,* p. 101; Otis T. Mason, *North American Bows, Arrows, and Quivers* (Washington, D.C., 1894), pp. 638, 640, 642, 644, 646; and Saxton T. Pope, *Bows and Arrows* (Berkeley, 1962), p. 34. Pope describes the bow at the Peabody Museum. It is artifact #49340 in the museum collection.

21. See Saxton T. Pope, *Hunting With the Bow and Arrow* (New York, 1947), pp. 41–43; Mason, *North American Bows,* p. 648; Wood, *New England's Prospect,* p. 97; Rosier, "A True Relation," p. 119, and Pope, *Bows and Arrows,* pp. 5, 10–19, 31–34, 62–63.

22. See Marc Lescarbot, *The History of New France,* ed. by W.L. Grant (3 vols.; Toronto, 1907–1914), II, p. 335; Gardiner, "Relation of the Pequot Warres," p. 144; J.A. Lundy, "Arrow Wounds," *Quarterly Bulletin of the Archaeological Society of Virginia,* VI, No. 4 (June, 1952), pp. 13–15; and Pope, *Bows and Arrows,* pp. 34, 62, 80–82, plate 16.

23. See Williams, *Key,* pp. 159–160; Mason, *North American Bows, Arrows, and Quivers,* pp. 646, 654–655; Pope, *Hunting with the Bow and Arrow,* p. 22; Charles C. Abbott, *Primitive Industry* (Salem, MA., 1881), p. 468; and Driver, *Indians of North America,* p. 183.

24. See Johnson, *Wonder-Working Providence,* p. 42; Pring, "A Voyage," pp. 56–57; Rosier, "A True Relation," p. 119; Verrazano, "Narrative," p. 18; Wood, *New England's Prospect,* pp. 18, 101; Morton, *New English Canaan,* pp. 186, 186fn.; Lescarbot, *The History of New France,* p. 191; Vaughan, *New England Frontier,* p. 38. Vaughan says arrows were eighteen inches long. This figure makes little sense from an archer's viewpoint and is refuted by the sources.

25. See Verrazano, "Narrative," p. 18; Rosier, "A True Relation," p. 119; Willoughby, *Antiquities*, pp. 131–134, 219–222, 225; *A Journal of the Pilgrims at Plymouth: Mourt's Relation* (New York, 1963), p. 37; Higginson, *New England's Plantation*, p. C4ᵛ; Frederic H. Douglas, "Copper and the Indians," *Denver Art Museum Indian Leaflet Series*, No. 75–76 (Denver, 1936), pp. 98–101; H.M. Chapin, "Indian Implements Found in Rhode Island," *RIHC*, XVII (1924), pp. 105, 107; Mason, *North American Bows*, pp. 653–654; Champlain, *Works*, I, pp. 325–326; and Wood, *New England's Prospect*, p. 101. For modern archaeological conclusions see William A. Ritchie, *The Archaeology of Martha's Vineyard* (New York, 1969), p. 232; Duncan Richie, "Lithic, Bone, and Antler Artifacts," in Gibson, ed., *Burr's Hill*, pp. 34, 37, 46–49; and Ripley P. Bullen, "Excavations in Northeastern Massachusetts," *Papers of the Robert S. Peabody Foundation for Archaeology*, I, No. 3 (Andover, MA., 1949), pp. 75, 78, 81–82, 87–88, 121–122, 138.

26. See Richie, "Lithic, Bone, and Antler Artifacts," pp. 34, 37; Russell, *Indian New England*, pp. 192–193; Pope, *Hunting with the Bow and Arrow*, pp. 22–24; F. Clark Howell, *Early Man* (New York, 1965), pp. 110–111; Frank Hamilton Cushing, "The Arrow," *American Anthropologist*, VIII (1895), pp. 316–319; Bullen, "Excavations," pp. 75, 138; Willoughby, *Antiquities*, pp. 222, 224; Otis T. Mason, *The Origins of Invention* (Cambridge, MA., 1966), pp. 129–130, 134–136; Mason, *North American Bows*, pp. 654–659; and Abbott, *Primitive Industry*, p. 468. Kenneth L. Feder, in "Of Stone and Metal: Trade and Warfare in Southern New England," *The New England Social Studies Bulletin*, vol. XLIV, no. 1(1986), pp. 27–30, concludes that lithic materials from the Hudson River Valley were common in Connecticut before the contact period, but that deteriorating relations between the Indians of the two regions cut off this trade.

27. See John Brereton, "A Briefe and True Relation of the Discoverie of the North Part of Virginia," *MHC*, VII, 3rd series (Boston, 1843), pp. 88, 91, 94; Douglas, "Copper and the Indians," pp. 98–99, 101; H.L. Reynolds, "Algonkin Metal-Smiths," *American Anthropologist*, I (1888), pp. 343–344; Roy Ward Drier and Octave Joseph Du Temple, eds., *Prehistoric Copper Mining in the Lake Superior Region* (private printing, 1961), pp. 19, 27, 31; Howe, *Prologue to New England*, pp. 20–21; Champlain, *Works*, I, pp. 164, 181–182, 184–185, 263; Higginson, *New England's Plantation*, p. C4ʳ; *Mourt's Relation*, p. 37; Harold A. Innis, *The Fur Trade in Canada* (New Haven, 1930), p. 11; Wood, *New England's Prospect*, p. 101; and Willoughby, *Antiquities*, pp. 235, 237.

28. See Champlain, *Works*, I, p. 326; Rosier, "A True Relation," p. 119; *Mourt's Relation*, p. 37; Higginson, *New England's Plantation*,

C4^r; Abbott, *Primitive Industry*, p. 205; Mason, *North American Bows*, p. 654; and Willoughby, *Antiquities*, pp. 219–222, 225.

29. See Williams, *Key*, p. 75; Pring, "A Voyage," p. 57; Morton, *New English Canaan*, p. 144; and Johnson, *Wonder-Working Providence*, p. 42.

30. Johnson, *Wonder-Working Providence*, p. 42.

31. See Wood, *New England's Prospect*, pp. 98–99; Pope, *Hunting With the Bow and Arrow*, pp. 26–27; John Metschl, "The Ralph Nunnemacher Collection of Projectile Arms," in *Bulletin of the Public Museum of the City of Milwaukee*, IX (1928), p. 22; Williams, *Key*, pp. 188–189; John Underhill, "Newes from America," *MHC*, VI, 3rd series (Boston, 1837), p. 26; and Nicolas Denys, *The Description and Natural History of the Coasts of North America (Acadia)*, ed. by William F. Ganong (Toronto, 1908), pp. 426–427.

32. See Wood, *New England's Prospect*, pp. 97–99; Thomas Lechford, *Newes from New England*, ed. by J. Hammond Trumbull (Boston, 1867), p. 120; Mason, *North American Bows*, pp. 647, 649; Metschl, "Nunnemacher Collection," p. 22; Pope, *Hunting With the Bow and Arrow*, pp. 26–27; Denys, *Description of North America*, pp. 426–427; and Johnson, *Wonder-Working Providence*, pp. 262–263.

33. See Josselyn, "An Account of Two Voyages," pp. 272, 309; Williams, *Key*, p. 156; Champlain, *Works*, I, pp. 338, 357, II, p. 96; Lescarbot, *History of New France*, II, p. 333; Willoughby, *Antiquities*, pp. 136–137, 141–144; Mason, *Origins of Invention*, pp. 141–142, 149; Chapin, "Indian Implements," pp. 110–111; Abbott, *Primitive Industry*, pp. 5–6; and Maurice Robbins, *The Amateur Archaeologist's Handbook* (New York, 1965), p. 70.

34. See Gookin, "Historical Collections," p. 152; Williams, *Key*, p. 156; Robbins, *Amateur Archaeologist's Handbook*, p. 175; Abbott, *Primitive Industry*, p. 38; Willoughby, *Antiquities*, pp. 142–144; Harold L. Peterson, *American Indian Tomahawks*, Contributions from the Museum of the American Indian Heye Foundation, XIX (New York, 1965), fig. 8, 9, 11, p. 86; Chapin, "Indian Implements," p. 110; and Charles C. Miles, *Indian and Eskimo Artifacts of North America* (Chicago, 1963), p. 384.

35. See Willoughby, *Antiquities*, pp. 136–137, 143–144; Abbott, *Primitive Industry*, pp. 10–11; Chapin, "Indian Implements," pp. 110–111; Peterson, *American Indian Tomahawks*, fig. 12–13, p. 87; and Miles, *Indian and Eskimo Artifacts*, pp. 82–83.

36. See Wood, *New England's Prospect*, p. 66; Josselyn, "An Account of Two Voyages," p. 309; Increase Mather, *Early History of New England*, ed. by Samuel G. Drake (Boston, 1864), p. 182; and Peterson,

American Indian Tomahawks, pp. 4–5, 8–9, 85, fig. 1.

37. See Wood, *New England's Prospect*, p. 95; Champlain, *Works*, I, pp. 357, 358; Willoughby, *Antiquities*, pp. 130–131; Denys, *Description of North America*, pp. 419–420; Josselyn, "An Account of Two Voyages," pp. 302–303; and Rosier, "A True Relation," p. 119.

38. See William Morrell, "Morrell's Poem of New England," p. 132; Champlain, *Works*, I, p. 418, II, pp. 88–89; Lechford, *Newes from New England*, p. 120; Underhill, "Newes from America," pp. 4–5; Wood, *New England's Prospect*, p. 95; and Winslow, "Good Newes From New England," pp. 288, 359–360.

39. See Rowlandson, "The Captivity," pp. 152–153; Williams, *Key*, pp. 126–128; Josselyn, "An Account of Two Voyages," p. 301; Morton, *New English Canaan*, pp. 149–151; Winslow, "Good Newes from New England," pp. 355–357, 366; Wood, *New England's Prospect*, pp. 92–94; and Champlain, *Works*, I, p. 417, II, pp. 87–88.

40. See Wood, *New England's Prospect*, p. 95; Gookin, "Historical Collections," p. 153; Church, *History of King Philip's War*, pp. 8–9; and Williams, *Key*, p. 191.

41. See Harry Holbert Turney-High, *The Practice of Primitive War*, (Missoula, MT., 1942), pp. 33, 99; and Vaughan, *New England Frontier*, pp. 39–40. Vaughan says, "European chroniclers who observed war among the Indians were unanimous in their contempt for the lackadaisical method with which it was conducted," but he does not explain the sophistication of much of the Indians' forest warfare.

42. See Turney-High, *Practice of Primitive War*, pp. 85–86; Wood, *New England's Prospect*, p. 95; Underhill, "Newes from America," pp. 10, 26; Mason, "A Brief History of the Pequot War," p. 149; Gookin, "Historical Collections," p. 164; and William Hubbard, *The History of the Indian Wars in New England*, ed. by Samuel G. Drake (2 vols.; Roxbury, MA., 1865), I, p. 113.

43. See Turney-High, *Practice of Primitive War*, pp. 33, 99; B.H. Quain, "The Iroquois," in *Cooperation and Competition Among Primitive Peoples*, ed. by Margaret Mead (Boston, 1961), p. 254; Williams, *Key*, p. 188; and Morton, *New English Canaan*, p. 248.

44. See Lescarbot, *History of New France*, III, p. 264; Gookin, "Historical Collections," pp. 64, 167; Josselyn, "An Account of Two Voyages," p. 309; Turney–High, *Practice of Primitive War*, p. 106; and *Winthrop Papers 1498-1649* (5 vols.; Boston, 1929–1947), V, p. 19.

45. See Turney-High, *Practice of Primitive War*, pp. 98–99; Wood, *New England's Prospect*, pp. 66–67; Axtell, "The Scholastic Philosophy of the Wilderness," pp. 342–43; and Champlain, *Works*, II, p. 85.

46. See Champlain, *Works,* II, pp. 85–86; Josselyn, "An Account of

Two Voyages," p. 309; Gookin, "Historical Collections," p. 166; Williams, *Key*, p. 11; Wood, *New England's Prospect*, p. 67; and Lescarbot, *History of New France*, III, p. 267.

47. Champlain, *Works*, II, pp. 88–89. In the actual French, Champlain refers to all men in authority here as "chefs." The translator tried to make the meaning clearer.

48. *Winthrop Papers*, III, pp. 413–414.

49. See Winslow, "Good Newes from New England," p. 366; *Winthrop Papers*, III, pp. 413–414; and Lescarbot, *History of New France* III, p. 264.

50. See Underhill, "Newes from America," p. 26; C.V. Wedgwood, *The Thirty Years War*, (New York, 1961), pp. 14–15, 86–87, 120–125, 248–250, 398–399; Hirsch, "The Collision of Military Cultures," pp. 1191–1194; and John Ferling, "The New England Soldier: A Study in Changing Perceptions," *American Quarterly*, XXXIII, No. 1 (Spring, 1981), pp. 27, 32–33.

51. Underhill, "Newes from America," p. 26.

52. Wood, *New England's Prospect*, p. 95.

53. See Mason, "Brief History of the Pequot War," p. 149; and Salisbury, *Manitou and Providence*, pp. 210, 215, 263fn.

54. Williams, *Key*, pp. 188–189.

55. See Hirsch, "The Collision of Military Cultures," pp. 1191–1194, 1199–1200; Turney-High, *Practice of Primitive War*, pp. 85–86; Wood, *New England's Prospect*, p. 95; Underhill, "Newes from America," pp. 10, 26; Mason, "A Brief History of the Pequot War," p. 149; Gookin, "Historical Collections," p. 164; and Hubbard, *History of the Indian Wars*, I, p. 113.

56. See Josselyn, "An Account of Two Voyages," p. 309; and Gookin, "Historical Collections," p. 164.

NOTES TO CHAPTER II

1. See T.J. Brasser, "Early Indian-European Contacts," in Bruce Trigger, ed., *Northeast*, vol. 15 of William C. Sturtevant, ed., *Handbook of North American Indians* (Washington, D.C., 1978–1984), pp. 78–88; Giovanni da Verrazano, "Narrative," in *Sailors' Narratives of Voyages Along the New England Coast 1524–1624*, ed. by George Parker Winship (Boston, 1905), pp. 16, 22; John Brereton, "A Briefe and True Relation of the Discoverie of the North Part of Virginia," *MHC*, VIII, 3rd series (Boston, 1843), pp. 85–86; Harold Innis, *The Cod Fisheries* (New Haven, 1940), pp. 11, 14–15, 26, 31, 71–72; Ralph Greenlee Lounsbury, *The British Fisherie at Newfoundland 1624–1763* (New Haven, 1934),

pp. 19–22; and Henry F. Howe, *Prologue to New England* (New York, 1943), pp. 9–11, 27, 47–50, 55–57, 85–86, 225. Howe thinks many fishermen visited New England in the sixteenth century, but this seems unlikely because of the dangers of approaching the coast and because good fishing there was evidently unknown before the seventeenth century. Howe does admit that there are no conclusive records of fishing voyages to the region and that southern New England was visited very rarely.

2. See Howe, *Prologue to New England*, pp. 3–4, 97, 162–165; Innis, *The Cod Fisheries*, pp. 71–73; Brereton, "A Briefe and True Relation," pp. 86, 97; Martin Pring, "A Voyage Set Out from the City of Bristoll" in *Sailors' Narratives of Voyages Along the New England Coast 1524–1624*, ed. by George Parker Winship (Boston, 1905), pp. 60–61; and James Rosier, "A True Relation of the Most Prosperous Voyage Made this Present Yeare 1605, by Captain George Waymouth in the Discovery of the Land of Virginia" in *Sailors' Narratives of Voyages Along the New England Coast 1524–1624*, ed. by George Parker Winship (Boston, 1905), pp. 131, 145–146.

3. See Brereton, "A Briefe and True Relation," pp. 85–86, 88, 91; and Gabriel Archer, "The Relation of Captain Gosnold's Voyage," *MHC*, VIII, 3rd series (Boston, 1843), pp. 73–75.

4. See William Bradford, *Of Plymouth Plantation 1620–1647*, ed. by S.E. Morison (New York, 1967), pp. 62, 69–70, 82–84; *A Journal of the Pilgrims at Plymouth: Mourt's Relation* (New York, 1963), pp. 31, 35–37; *The Works of Samuel de Champlain*, ed. by H.P. Biggar (6 vols.; Toronto, 1922–36), I, pp. 353–354, 419–420, 429–430; John Smith, *Travels and Works*, ed. by Edward Arber (2 vols.; Edinburgh, 1910), II, p. 719; William Wood, *New England's Prospect* (Boston, 1865), pp. 85–86; Marc Lescarbot, *The History of New France* (3 vols.; Toronto, 1907–1914), II, pp. 332, 334, 337; Rosier, "A True Relation," p. 115; Verrazano, "Narrative," pp. 21–22; Howe, *Prologue to New England*, pp. 12–13; Alden Vaughan, *New England Frontier: Puritans and Indians 1620–1675* (Boston, 1965), pp. 6–19; and Carolyn Thomas Foreman, *Indians Abroad 1492–1938* (Norman, OK., 1943), pp. 9, 15–21.

5. See Thomas Dermer, "Narrative," in *Sailors' Narratives of Voyages Along the New England Coast 1524–1624*, ed. by George Parker Winship (Boston, 1905), p. 251; John Josselyn, "An Account of Two Voyages to New England," *MHC*, III, 3rd series (Cambridge, MA., 1833), pp. 293–294; Vaughan, *New England Frontier*, pp. 21–22, 28; Daniel Gookin, "Historical Collections of the Indians in New England," *MHC*, I, 1st series (Boston, 1792), p. 18; Howe, *Prologue to New England*, pp. 280–282, 287; and Neal Salisbury, *Manitou and Provi-*

dence: *Indians, Europeans, and the Making of New England, 1500–1643* (New York, 1982), pp. 101–105.

6. See Bradford, *Of Plymouth Plantation*, p. 87; Robert Cushman, "A Sermon Preached at Plymouth" in *Chronicles of the Pilgrim Fathers of the Colony of Plymouth from 1602 to 1625*, ed. by Alexander Young (Boston, 1841), pp. 258–259; Edward Winslow, "Good Newes from New England," in *Chronicles of the Pilgrim Fathers of the Colony of Plymouth from 1602 to 1625*, ed. by Alexander Young (Boston, 1841), p. 359; Thomas Morton, *New English Canaan* (New York, 1967), pp. 132–134; Howe, *Prologue to New England*, pp. 280–282, 287; Vaughan, *New England Frontier*, pp. 21–22; and Salisbury, *Manitou and Providence*, pp. 104–106.

7. See Bradford, *Of Plymouth Plantation*, pp. 260, 270–271; Josselyn, "An Account of Two Voyages," pp. 298–300; Vaughan, *New England Frontier*, p. 103; and Salisbury, *Manitou and Providence*, pp. 190–191, 209–210.

8. See James Axtell, *The European and the Indian: Essays in the Ethnohistory of Colonial North America* (Oxford, 1981), pp. 74, 115–122; Peter Thomas, "In the Maelstrom of Change: The Indian Trade and Cultural Process in the Middle Connecticut River Valley, 1635–1665" (Ph.D. dissertation, University of Massachusetts, 1979), pp. 11–14; Paul Robinson, Marc A. Kelley, and Patricia Rubertone, "Preliminary Biocultural Interpretations from a Seventeenth-Century Narragansett Indian Cemetery in Rhode Island," in William W. Fitzhugh, ed., *Cultures in Contact: The Impact of European Contacts on Native American Cultural Institutions A.D. 1000–1800* (Washington, 1985), pp. 119–124; Susan G. Gibson, ed., *Burr's Hill* (Providence, R.I., 1980), pp. 19–21; Christina B. Johannsen, "European Trade Goods and Wampanoag Culture in the Seventeenth Century," in *Burr's Hill*, ed. by Gibson, pp. 25–33; William S. Simmons, *Cautantowwit's House* (Providence, R.I., 1970), pp. 68, 159–160; William Morrell, "Morrell's Poem of New England," *MHC*, I, 1st series (Boston, 1792), p. 134; Wood, *New England's Prospect*, p. 105; Increase Mather, *Early History of New England*, ed. by Samuel G. Drake (Boston, 1864), pp. 289–292; Roger Williams, *A Key Into the Langue of America* (Providence, R.I., 1936), p. 203; and Nicholas Denys, *The Description and Natural History of the Coasts of North America (Acadia)*, ed. by William F. Ganong (Toronto, 1908), pp. 439–443.

9. See Patricia E. Rubertone, "Archaeology, colonialism and 17th-century Native America: towards an alternative interpretation," in R. Layton, ed., *Conflict in the Archaeology of Living Traditions* (London, 1989), pp. 36–37; Robinson, Kelley, and Rubertone, "Preliminary Biocultural Interpretations," p. 125; James W. Bradley, *Evolution of the*

Onondaga Iroquois: Accommodating Change, 1500–1655 (Syracuse, 1987), pp. 6, 80, 131–133, 146–147, 171–176; Nancy Lurie, "Indian Cultural Adjustment to European Civilization," in *Seventeenth-Century America,* ed. by James Morton Smith (Chapel Hill, 1959), p. 38; George Irving Quimby, *Indian Culture and European Trade Goods* (Madison, 1966), pp. 8–11; H.G. Barnett, "Invention and Cultural Change," *American Anthropologist,* XLIV (January–March, 1942), pp. 23–26; and H.G. Barnett, "Acculturation: An Exploratory Formulation," Social Science Research Council Summer Seminar on Acculturation, 1953, in *American Anthropologist,* LXVI (1954), pp. 973–1002. The author has benefitted greatly from discussions with Patricia Rubertone, of Brown University, and Paul Robinson, of the Rhode Island Historical Preservation Commission, who let him see a number of artifacts recovered from Narragansett burials at the RI 1000 site. He is also grateful to Robert B. Gordon, of Yale University, for reviewing a metallurgical report on RI 1000 artifacts and offering his insights on the reworking of iron objects.

10. See Johannsen, "European Trade Goods," p. 28; Lisa Fiore, "Implements of Iron," in Gibson, ed., *Burr's Hill,* p. 96; Harold L. Peterson, *American Indian Tomahawks,* Contributions From the Museum of the American Indian, Heye Foundation XIX (New York, 1965), pp. 11, 14–15, 19–20; and Quimby, *Indian Culture and European Trade Goods,* pp. 8–11.

11. See Daniel Gookin, "Historical Collections," p. 152; Champlain, *Works,* I, pp. 338, 417; Wood, *New England's Prospect,* pp. 87–88; Peterson, *American Indian Tomahawks,* pp. 11, 14–15, 19–20; Simmons, *Cautantowwit's House,* pp. 100, 104, 152, 158–160; and Johannsen, "European Trade Goods," p. 30. Steel knives may have made it easier to take scalps. Apparently more heads were taken as trophies than scalps in New England aboriginal warfare (it may have been quicker to cut off an entire head with a stone axe than to strip a scalp during a battle), but scalping was preferred when Indians had to carry their trophies back from a great distance. James Axtell, in *The European and the Indian: Essays in the Ethnohistory of Colonial North America* (Oxford, 1981), pp. 144, 209, 213–214, proves that scalping was not introduced in the New World by Europeans. His volume has much to offer on technological diffusion and military systems in New England. See also Lorraine F. Williams, "Ft. Shantok and Ft. Corchaug: A Comparative Study of Seventeenth-Century Culture Contact in the Long Island Sound Area" (Ph.D. dissertation, New York University, 1972), pp. 248–251, 259; Robinson, Kelley, and Rubertone, "Preliminary Biocultural Interpretations," p. 120; and Bradley, *Evolution of the Onondaga Iroquois,* pp. 146–147. Bradley suggests that even new hatchet or axe heads were cut up to make other important tools. However, broken or otherwise

damaged specimens would have been the first candidates for this conversion. Various examples of wedges or chisels that appear to have been made from hatchet heads have been found at Fort Shantok and at the RI 1000 site.

12. My conclusions about the bow and the muzzle-loading firearm are based on personal experience in hunting with both types of weapons.

13. See Williams, *Key*, p. 84; Lurie, "Indian Cultural Adjustment," p. 38; Peter Farb, *Man's Rise to Civilization as Shown by the Indians of North America from Primeval Times to the Coming of the Industrial State* (New York, 1968), pp. 9–10, 22; Rosier, "A True Relation," p. 115; and Wood, *New England's Prospect*, pp. 85, 87–88.

14. Bradford, *Of Plymouth Plantation*, p. 207.

15. See Edward Arber, ed., *The Story of the Pilgrim Fathers 1606–1623 A.D.* (Boston, 1897), p. 338; *Mourt's Relation*, p. 23; MCR, I, pp. 25–26; Harold L. Peterson, "The Military Equipment of the Plymouth and Bay Colonies 1620–1690," *New England Quarterly*, XX, no. 2 (1947), pp. 204–207; and Harold L. Peterson, *Arms and Armor in Colonial America 1526–1783* (Harrisburg, PA., 1956), p. 44. For broad coverage of colonial firearms, see M.L. Brown, *Firearms in Colonial America: The Impact on History and Technology 1492–1762* (Washington, D.C., 1980).

16. See Bernard and Fawn Brodie, *From Crossbow to H-Bomb* (New York, 1962), pp. 80–81; Peterson, *Arms and Armor*, pp. 14, 15, 44; and Howard L. Blackmore, *Firearms* (New York, 1964), p. 12.

17. See Theodore Ropp, *War in the Modern World* (New York, 1962), pp. 32, 51; Brodie, *Crossbow to H-Bomb*, pp. 80-81; B.H. Liddell Hart, "Armed Forces and the Art of War: Armies" in *The Zenith of European Power 1830–1870, The New Cambridge Modern History*, X (Cambridge, Eng., 1967), p. 303; Francis Grose, *Military Antiquities* (2 vols.; London, 1812), II, p. 124.

18. See John Metschl, "The Rudolph J. Nunnemacher Collection of Projectile Arms," *Bulletin of the Public Museum of the City of Milwaukee*, IV (1928), p. 55; Grose, *Military Antiquities*, II, p. 127; and Peterson, *Arms and Armor*, p. 55.

19. See Grose, *Military Antiquities*, I, p. 153 fn., II, pp. 127, 295; and *Mourt's Relation*, p. 23.

20. Howe, *Prologue to New England*, pp. 159–160.

21. See Peterson, *Arms and Armor*, pp. 15–17, 19–20; Peterson, "Military Equipment," pp. 202–203; William Bariffe, *Military Discipline or the Young Artillery-Man* (London, 1643), pp. 3–5; and Richard Elton, *The Compleat Body of the Art Military* (London, 1659), pp. 5–6.

22. *Mourt's Relation*, p. 35.

23. Peterson, *Arms and Armor*, pp. 25–36.

24. *Ibid.*, pp. 14–17, 26–27.

25. See Elton, *The Compleat Body of the Art Military*, pp. 5–6; Bariffe, *Military Discipline*, pp. 3–6; C.H. Firth, *Cromwell's Army* (London, 1962), pp. 94–98; Allen French, "The Arms and Military Training of Our Colonizing Ancestors," *Massachusetts Historical Society Proceedings*, LXVII (Boston, 1945), pp. 18–19; Ropp, *War in the Modern World*, pp. 50–51; and Peterson, *Arms and Armor*, pp. 17–22, 160.

26. See "Records of the Company of the Massachusetts Bay to the Embarkation of Winthrop and His Associates for New England," *Transactions and Collections of the American Antiquarian Society*, III (1857), p. 87; and *Mourt's Relation*, p. 35.

27. Underhill, "Newes from America," p. 23.

28. See Bradford, *Of Plymouth Plantation*, p. 342; and William Brigham, ed., *The Compact With the Charter and Laws of the Colony of New Plymouth* (Boston, 1836), pp. 84, 184.

29. See *NHCR*, I, pp. 96–97, 131, 503; *CCR*, I, pp. 15, 542–544, II, pp. 285, 383; *MCR*, I, pp. 25–26, 84–85, II, p. 122, III, pp. 169, 368, IV–1, p. 6, IV–2, p. 562, V, pp. 47–48, 51, 63; MA, LXVIII, pp. 135a, 1436; Edward Randolph, *Edward Randolph 1676–1703*, ed. by Robert Toppan (2 vols.; New York, 1967), II, pp. 207, 238; Gookin, "Historical Collections," p. 509; Peterson, "Military Equipment," pp. 204–205; Sylvester Judd, *History of Hadley* (Springfield, MA., 1905), pp. 220–221; Richard LeBaron Bowen, *Early Rehoboth* (4 vols.; Concord, N.H., 1948), III, pp. 60–61; James Savage, "Gleanings for New England History," *MHC*, VIII, 3rd series (Boston, 1842), p. 333; and *Acts and Resolves Public and Private of the Province of the Massachusetts Bay* (Boston, 1869), I, p. 129.

30. See Patrick M. Malone, "Indian and English Military Systems in New England in the Seventeenth Century," (Ph.D. Diss., Brown Univ., 1971), pp. 76–93, 196–199; Carl P. Russell, *Guns on the Early Frontiers* (Berkeley, 1957), pp. 16–26, 40–42; Judd, *History of Hadley*, p. 183; Mather, *Early History of New England*, p. 206; CA, War Council I, p. 44; Simmons, *Cautantowwit's House*, pp. 82–84; Peterson, *Arms and Armor*, pp. 14–22, 23–35, 43–45; Joseph R. Mayer, *Flintlocks of the Iroquois 1620–1687*, Research Records of Rochester Museum of Arts and Sciences, no. 6 (Rochester, 1943), pp. 31, 32–33; and T.M. Hamilton, "Some Gun Parts from 17th Century Seneca Sites," in *Indian Trade Guns*, ed. by T.M. Hamilton, *The Missouri Archaeologist*, XXII (Columbia, MO., 1960), pp. 99–119. Mayer and Hamilton provide extensive archaeological evidence that Iroquoian Indians in New York switched rapidly from bows to flintlocks in the 17th century and rarely used matchlocks.

31. See Champlain, *Works*, I, p. 338; Lechford, *Newes from New England*, p. 119; Robert Juet, "Narrative of Henry Hudson's Voyage" in *Sailors' Narratives of Voyages Along the New England Coast 1524–1624*, ed. by George Parker Winship (Boston, 1905), p. 182; Gookin, "Historical Collections," pp. 156–157; Innis, *The Fur Trade in Canada*, pp. 6–8; Calvin Martin, *Keepers of the Game* (Berkeley, CA., 1978), pp. 61–62, 153–155; and L.E. Babits, "The Evolution and Adoption of Firearm Ignition Systems in Eastern North America: An Ethnohistorical Approach," *The Chesopiean*, vol. 14, nos. 3–4 (1976), pp. 61–65.

32. See Denys, *Description of North America*, pp. 399, 440–443; Bradford, *Of Plymouth Plantation*, pp. 194, 204; Samuel Eliot Morison, *Builders of the Bay Colony* (Cambridge, MA., 1958), pp. 341–343; Gookin, "Historical Collections," pp. 156–157; Vaughan, *New England Frontier*, pp. 213–220; and Innis, *The Fur Trade in Canada*, pp. 12–13.

33. See Neal Salisbury, *Manitou and Providence*, pp 148–152; Neal Salisbury, "Toward the Covenant Chain: Iroquois and Southern New England Algonquians, 1637–1684," in Daniel Richter and James H. Merrell, eds., *Beyond the Covenant Chain: The Iroquois and Their Neighbors in Indian North America, 1600–1800* (Syracuse, 1987), pp. 61–63; Patricia E. Rubertone, "Archaeology, colonialism and 17th-century Native America," pp. 39–43; Katherine Billings, "Beads of Shell and Glass," in Gibson, ed., *Burr's Hill*, pp. 118–119; Bradford, *Of Plymouth Plantation*, pp. 203–204; Wood, *New England's Prospect*, p. 69; Williams, *Key*, pp. 152–153; Josselyn, "An Account of Two Voyages," pp. 306–307; Lescarbot, *History of New France*, III, pp. 158, 158fn.; Gookin, "Historical Collections," p. 152; Alfred Goldsworthy Bailey, "The Conflict of European and Eastern Algonkin Cultures, 1504–1700" (Ph.D. Diss., University of Toronto, 1934), pp. 259–260; Regina Flannery, *An Analysis of Coastal Algonquian Culture*, Catholic University of American Anthropological Series, No. 7 (Washington, D.C., 1939), pp. 119–121; Willoughby, *Antiquities*, pp. 266–267, 270–271; Vaughan, *New England Frontier*, pp. 221–223; Morison, *Builders of the Bay Colony*, pp. 341, 346, 354–355; and CCR, I, pp. 13, 61, 179.

34. Morton, *New English Canaan*, pp. 157–158.

35. See Bradford, *Of Plymouth Plantation*, pp. 203–204; and Wood, *New England's Prospect*, p. 69. For discussion of the ways in which trade patterns created both interdependency and conflict, see Salisbury, *Manitou and Providence*, pp. 151–152; Neal Salisbury, "Social Relationships on a Moving Frontier: Natives and Settlers in Southern New England, 1638–1675," *Man in the Northeast*, no. 33 (1987), pp. 90–93; Salisbury, "Beyond the Covenant Chain," pp. 62–67; and Thomas, "In

the Maelstrom of Change," pp. 14–16, 180–182, 196–197.

36. See Denys, *Description of North America*, pp. 429–432, 443; Wood, *New England's Prospect*, pp. 29, 99–100; David Thompson, *Narrative of His Explorations in Western America 1784–1812*, ed. by J.B. Tyrrell (Toronto, 1916), pp. 112, 200–201, 204–206; and Innis, *The Fur Trade in Canada*, pp. 25, 266.

37. See Underhill, "Newes from America," p. 17; Isaack de Rasieres, "Letter of Isaack de Rasieres to Samuel Blommaert, 1628(?)" in *Narratives of New Netherland 1609–1664*, ed. by J. Franklin Jameson (New York, 1959), p. 229; Brigham, *Laws of New Plymouth* (Boston, 1836), p. 94; Williams, *Key*, p. 112; Bradford, *Of Plymouth Plantation*, pp. 204, 206–208; Denys, *Description of North America*, p. 399; Innis, *The Fur Trade in Canada*, p. 31; Gookin, "Historical Collections," p. 152; John Pory, "John Pory to the Governor of Virginia (Sir Francis Wyatt)," *Three Visitors to Earl Plymouth*, ed. by Sydney V. James, Jr. (Plymouth, MA., 1963), p. 16; and *CCR*, I, pp. 77–79.

38. See Gookin, "Historic Collections," p. 152; Morison, *Builders of the Bay Colony*, pp. 341, 345, 351–352, 358; *MCR*, I, p. 48, II, pp. 4, 61, III, p. 65, IV–1, p. 291, IV–2, p. 2; *CCR*, I, pp. 2, II, 13, 74, 138; Morton, *New English Canaan*, pp. 157–158; Bradford, *Of Plymouth Plantation*, pp. 203–204, 207; Hubbard, *The History of the Indian Wars in New England*, I, pp. 56–57; William Bradford, "Letter Book," *MHC*, III, 1st series (Boston, 1810), p. 63; and John and William Pynchon, "The Pynchon Court Records," in *Colonial Justice in Western Massachusetts (1639–1702)*, ed. by Joseph H. Smith (Cambridge, MA., 1961), p. 208.

39. Vaughan, *New England Frontier*, p. 22.

40. See Hunt, *The Wars of the Iroquois*, pp. 18–22; and Gookin, "Historical Collections," p. 161.

41. See William Harris, *A Rhode Islander Reports on King Philip's War: The Second William Harris Letter of August 1676*, ed. by Douglas Edward Leach (Providence, 1963), pp. 53–57; *MCR*, III, pp. 38–39, IV–2, p. 23; *PCR*, IX, pp. 10–12; Edward Winslow, *Hypocrisie Unmasked* (Providence, 1916), p. 86; John Winthrop, *A Declaration of Former Passages and Proceedings Betwixt the English and Narrowgansetts* (Cambridge, MA., 1645), pp. 2–7; Bradford, *Of Plymouth Plantation*, p. 330; and Douglas Edward Leach, *Flintlock and Tomahawk: New England in King Philip's War* (New York, 1966), pp. 22–23.

NOTES TO CHAPTER III

1. See Edward Winslow, *Hypocrisie Unmasked* (Providence, 1916), p. 82; Charles Francis Adams, "Introduction" in Thomas Morton,

New English Canaan (New York, 1967), pp. 20, 22; George T. Hunt, *The Wars of the Iroquois* (Madison, Wisconsin, 1960), pp. 170–175; William Bradford, *Of Plymouth Plantation 1620–1647*, ed. by S.E. Morison (New York, 1967), p. 204; William Bradford, "Letter Book," *MHC*, III, 1st series (Boston, 1810), pp. 63–64; John Pory, "John Pory to the Governor of Virginia (Sir Francis Wyatt)," in *Three Visitors to Early Plymouth*, ed. by Sydney V. James, Jr. (Plymouth, MA., 1963), p. 16; *MCR*, IV–1, p. 291; Daniel Gookin, "Historical Collections of the Indians in New England," *MHC*, I, 1st series (Boston, 1792), p. 152; William Wood, *New England's Prospect* (Boston, 1865), p. 67; Harold A. Innis, *The Fur Trade in Canada* (New Haven, 1930), pp. 18, 33, 53; and Allen W. Trelease, *Indian Affairs in Colonial New York: The Seventeenth Century* (Ithaca, New York, 1960), pp. 95–99, 136.

2. Bradford, *Of Plymouth Plantation*, pp. 203–204. See also Neal Salisbury, *Manitou and Providence: Indians, Europeans, and the Making of New England, 1500–1643* (New York, 1982), pp. 148–152, 156–158, 185; and Peter Thomas, "In the Maelstrom of Change: The Indian Trade and Cultural Process in the Middle Connecticut River Valley, 1635–1665" (Ph.D. dissertation, University of Massachusetts, 1979), pp. 180–182, 191–192, 196–197. Salisbury points out that Bradford exaggerated the extent of the early arming of Indians in his region. Bradford's hyperbolic outrage is, however, an indication of how serious he considered the offense of illegally arming Indians.

3. See Royal Proclamations of 1622 and 1630 in *Historical Collections*, ed. by Ebenezer Hazard (2 vols.; Philadelphia, 1792 and 1794), I, pp. 151, 311; Adams, "Introduction," pp. 20–21, 26; Bradford, *Of Plymouth Plantation*, pp. 207, 232fn.–233fn.; Bradford, "Letter Book," pp. 61–63; and *MCR*, I, p. 48.

4. See Bradford, *Of Plymouth Plantation*, pp. 206–207, 209–210, 216–217; Adams, "Introduction," pp. 21, 26–29, 35–38; and Neal Salisbury, *Manitou and Providence: Indians, Europeans, and the Making of New England, 1500–1643* (New York, 1982), p. 185. Salisbury refers to Bradford's "hysteria" and says that the claims about Morton were exaggerated.

5. See "Records of the Company of the Massachusetts Bay to the Embarkation of Winthrop and His Associates for New England," *Transactions and Collections of the American Antiquarian Society*, III (1857), p. 88; and *MCR*, I, pp. 99–100.

6. See *MCR*, I, pp. 99–100, 195–196, II, p. 16, III, pp. 193, 308, 402, IV–1, pp. 10–11, 133, 251, 352; John and William Pynchon, "The Pynchon Court Records," in *Colonial Justice in Western Massachusetts (1639–1702)*, ed. by Joseph H. Smith (Cambridge, MA., 1961), p. 208; MA, XXX, pp. 28, 63–64.

7. *NHCR*, I, p. 206. See *NHCR*, I, p. 60 for an earlier law on the subject.

8. See *CCR*, I, pp. 1–2, 49, 72, 74, 79, 80,106, 146, 163, 167, 171, 182, 198–199, 218–219, 242, 530–531; "Abstract of Laws," in Peter Force, ed., *Tracts and Other Papers Relating Principally to the Origin, Settlement, and Progress of the Colonies in North America* (4 vols.; Washington, D.C., 1836–1846), III, No. 9, p. 10; Bradford, *Of Plymouth Plantation*, pp. 207–208, 232–233; William Brigham, ed., *The Compact With the Charter and Laws of the Colony of New Plymouth* (Boston, 1836), pp. 65, 76; and *PCR*, II, pp. 8, 36, III, p. 6.

9. William Bradford, "A Descriptive and Historical Account of New England in Verse," *MHC*, III, 1st series (Boston, 1810), p. 83. See also Bradford, "Letter Book," p. 63 for a complaint in 1628 about arms trade under the pretense of fishing.

10. Roger Williams, *The Complete Writings of Roger Williams* (7 vols.; New York, 1963), VI, p. 296.

11. *MCR*, III, p. 402.

12. See Bradford, *Of Plymouth Plantation*, pp. 203–204, 275, 279; Roger Williams, *A Key Into the Language of America* (Providence, R.I., 1936), p. 185; Williams, *The Complete Writings*, VI, pp. 296, 303; *PCR*, I, pp. 22, 107, 209, X, 9–18, 62; *CCR*, I, pp. 113, 197; *MCR*, III, p. 208, IV–1, p. 21; Thomas Lechford, *Newes from New England*, ed. by J. Hammond Trumbull (Boston, 1867), p. 117; and Edward Johnson, *Johnson's Wonder-Working Providence*, ed. by J. Franklin Jameson (New York, 1910), p. 148. Bruce Trigger, "Early Iroquoian Contacts with Europeans," in Bruce Trigger, ed., *Northeast*, vol. XV of William C. Sturtevant, ed., *Handbook of North American Indians* (Washington, D.C., 1978–1984), pp. 352–354, suggests that the Iroquois were trading furs for guns in New England in 1640, and that the Dutch, at that point, had to relax their restraints on firearms in order to compete with the New England traders. See also Neal Salisbury, "Social Relationships on a Moving Frontier: Natives and Settlers in Southern New England, 1638–1675," *Man in the Northeast*, no. 33 (1987), pp. 90–91. Salisbury stresses the extensive trade relationships and interdependence between the English and the Indians in New England.

13. See Wood, *New England's Prospect*, p. 67; and Bradford, *Of Plymouth Plantation*, p. 204.

14. See Innis, *The Fur Trade*, pp. 52–53; Hunt, *The Wars of the Iroquois*, pp. 172–175; and Trelease, *Indian Affairs in Colonial New York*, pp. 217–218.

15. See Hunt, *The Wars of the Iroquois*, pp. 166–171; Carl P. Russell, *Guns on the Early Frontiers* (Berkeley, 1957), pp. 12–13; Trelease, *Indian Affairs in Colonial New York*, pp. 95–96, 136; J. Franklin

Jameson, ed., *Narratives of New Netherland 1609–1664* (New York, 1959), p. 274; PCR, X, p. 62; and Innis, *The Fur Trade*, p. 33.

16. See Johnson, *Wonder-Working Providence*, p. 148; Williams, *The Complete Writings*, VI, p. 42; PCR, IX, p. 107, X, pp. 3–26; and John Mason, "A Brief History of the Pequot War," MHC, VIII, 2nd series (Boston, 1819), p. 13.

17. See PCR, IX, pp. 3–8, 22, 107, 148–149, 172, 177–178, 191; CCR, I, pp. 113, 138, 145, 197; and MCR, III, p. 208.

18. PCR, IX, pp. 107, 176, X, pp. 17, 27–29, 59.

19. See PCR, IX, pp. 173, 176–177, X, pp. 17–18, 26–28, 61–62, 72; and NHCR, I, pp. 513, 522–524, 528–529.

20. See PCR, X, pp. 26–29, 30–34, 55–57, 62; MA, XXX, pp. 27, 27a; Trelease, *Indian Affairs in Colonial New York*, pp. 100, 106–107; and Hunt, *The Wars of the Iroquois*, pp. 170–171.

21. See Trelease, *Indian Affairs in Colonial New York*, pp. 190–191; Russell, *Guns on the Early Frontiers*, pp. 13–14; and MA, XXX, pp. 120, 120b, 122.

22. See MCR, II, pp. 44, 61, 163, 268, III, pp. 65, 164; PCR, III, p. 192, IX, pp. 105, 149; Brigham, ed., *Laws of New Plymouth*, p. 94; and MCR, II, p. 61.

23. MCR, II, p. 61.

24. PCR, IX, p. 105.

25. See MCR, II, p. 268, III, p. 164; and PCR, IX, p. 149.

26. Brigham, ed., *Laws of New Plymouth*, p. 94.

27. See MA, XXX, pp. 120, 120b, 122; and Neal Salisbury, "Toward the Covenant Chain: Iroquois and Southern New England Algonquians, 1637–1684," in Daniel Richter and James H. Merrell, eds., *Beyond the Covenant Chain: The Iroquois and Their Neighbors in Indian North America, 1600–1800* (Syracuse, 1987), p. 68.

28. Brigham, ed., *Laws of New Plymouth*, pp. 148, 152.

29. See MCR, IV-2, p. 365; and Brigham, ed., *Laws of New Plymouth*, p. 158.

30. See CCR, II, p. 119; Brigham, ed., *Laws of New Plymouth*, p. 158; and Alden Vaughan, *New England Frontier: Puritans and Indians 1620–1675* (Boston, 1965), pp. 229–230.

31. CCR, II, p. 119.

32. See PCR, V, pp. 63–64, 109; George D. Langdon, Jr., *Pilgrim Colony: A History of New Plymouth 1620–1691* (New Haven, 1966), pp. 159–161; Douglas Edward Leach, *Flintlock and Tomahawk: New England in King Philip's War* (New York, 1966), pp. 26–29; and Brigham, ed., *Laws of New Plymouth*, pp. 171, 288, 329, 311–333.

33. William Harris, *A Rhode Islander Reports on King Philip's War: The Second William Harris Letter of August 1676*, ed. by Douglas

Edward Leach (Providence, 1963), p. 23. See also "Testimony of Hugh Cole," at Plymouth, March 8, 1670/71, Misc. MSS, Mass. Historical Society; and Edward Randolph, *Edward Randolph 1676–1703*, ed. by Robert Toppan (2 vols.; New York, 1967), II, pp. 245–246 for a very biased attack on the arms trade begun by Massachusetts Bay.

34. See *MCR*, II, pp. 24, 27; John and William Pynchon, "The Pynchon Court Records," p. 263; *CCR*, I, pp. 240, 293, 375; CA, Indians, I, p. 14; and Brigham, ed., *Laws of New Plymouth*, pp. 100, 288, 329–331.

35. See *CCR*, I, pp. 293, 351, 375; and Brigham, ed., *Laws of New Plymouth*, p. 100.

36. See *MCR*, III, p. 397, IV–1, pp. 86, 257; Randolph, *Edward Randolph*, II, pp. 244–245.

37. See Bradford, *Of Plymouth Plantation*, p. 339; *Winthrop Papers*, 1498–1649 (5 vols.; Boston, 1929–1947), V, pp. 19, 85; and Increase Mather, *A Brief History of the War With the Indians in New England*, ed. by Samuel G. Drake (Albany, N.Y., 1862), p. 47.

38. Gookin, "Historical Collections," pp. 166–167.

NOTES TO CHAPTER IV

1. See Danby Pickering, ed., *Statutes At Large* (8 vols.; Cambridge, England, 1763), IV, pp. 163, 272–273, V, pp. 71–72, VI, pp. 212–215, 346–347, VII, pp. 134–141, 184, VIII, pp. 380–381; Carl Bridenbaugh, *Vexed and Troubled Englishmen 1590–1642* (New York, 1968), p. 379; E. Alexander Bergstrom, "English Game Laws and Colonial Food Shortages," *The New England Quarterly*, XII (March–December, 1939), pp. 685–688; Joseph Strutt, *The Sports and Pastimes of the People of England*, ed. by J. Charles Cox (London, 1903), pp. 5, 13–16, 18, 25; *Winthrop Papers*, 1498–1649 (5 vols.; Boston, 1929–1947), I, pp. 165, 307; and Gervase Markham, *County Contentments: Or, The Husbandmans Recreations* (London, 1633), pp. 32–35, 53–55. Markham covers hunting practices in detail but does not mention the use of firearms.

2. See "The King's Christian Duties," quoted in Strutt, *Sports and Pastimes*, p. 5; Pickering, ed., *Statutes*, IV, pp. 141, 161, V, pp. 71–72, VIII, pp. 380–381; Philip Alexander Bruce, *Social Life of Virginia in the Seventeenth Century* (Lynchburg, VA., 1927), p. 124; Bergstrom, "English Game Laws," pp. 688–689; and Lindsay Boynton, *The Elizabethan Militia 1558–1638* (Toronto, 1967), p. 119.

3. *Winthrop Papers*, I, p. 165.

4. See Bridenbaugh, *Vexed and Troubled Englishmen*, p. 379; and Bergstrom, "English Game Laws," p. 686.

5. See Edward Winslow, "A Letter" [Dec. 11, 1621] in *A Journal of the Pilgrims at Plymouth: Mourt's Relation* (New York, 1963), p. 86; and Bergstrom, "English Game Laws," pp. 688–689.

6. See Bergstrom, "English Game Laws," p. 686; Bridenbaugh, *Vexed and Troubled Englishmen*, pp. 95, 379; and *Winthrop Papers*, I, pp. 307–308.

7. See "Instructions for the Execution of the Commission . . . ," in Francis Grose, *Military Antiquities* (2 vols.; London, 1812), I, pp. 76, 83–84, 87; Thomas Garden Barnes, *Somerset 1625–1640* (Cambridge, MA., 1961), p. 107; C.H. Firth, *Cromwell's Army* (London, 1962), p. 5; and Boynton, *Elizabethan Militia*, pp. 7, 91–92.

8. C.G. Cruickshank, *Elizabeth's Army* (Oxford, 1946), p. 8.

9. See Boynton, *Elizabethan Militia*, pp. 112–113, 257; Firth, *Cromwell's Army*, pp. 7–8; Grose, *Military Antiquities*, I, pp. 85, 135, 333–334; Henry J. Webb, *Elizabethan Military Science* (Madison, Wisconsin, 1965), pp. 92–98; J.W. Fortescue, *A History of the British Army* (13 vols.; London, 1910–1930), I, pp. 136, 138–139; "A Village Armoury," *East Anglian Miscellany*, reference no. 712 (1965), pp. 32–33.

10. See Boynton, *Elizabethan Militia*, pp. 113–115, 117–119; Grose, *Military Antiquities*, I, pp. 76, 85; Firth, *Cromwell's Army*, pp. 3, 5, 9–10; and Barnes, *Somerset*, pp. 107–108.

11. See Barnes, *Somerset*, pp. 107, 109, 112, 123, 248; Firth, *Cromwell's Army*, pp. 5–6, 11; and Boynton, *Elizabethan Militia*, pp. 209–210, 216, 237–242, 244–245.

12. For detailed studies of the militia under Charles I, see Barnes, *Somerset*, pp. 106–123, 244–280; and Boynton, *Elizabethan Militia*, pp. 244–297.

13. See Barnes, *Somerset*, pp. 244–280; and Boynton, *Elizabethan Militia*, pp. 244–297.

14. See Boynton, *Elizabethan Militia*, pp. 238–240, 244, 246–249, 254, 257–260, 266–269; Barnes, *Somerset*, pp. 109, 113–114, 117–119, 248–250; Grose, *Military Antiquities*, II, pp. 323–337; Firth, *Cromwell's Army*, pp. 94–98; *Instructions for Musters and Armes, and the Use Thereof: By Order from the Lords of His Majesties Most Honourable Privy Counsayle* (London, 1623), pp. 1–12; and see William Bariffe, *Military Discipline or the Young Artillery-Man* (London, 1643) for very complicated military techniques from Europe.

15. See Richard Elton, *The Compleat Body of the Art Military* (London, 1659), pp. 5–6; Bariffe, *Military Discipline*, pp. 3–6; Firth, *Cromwell's Army*, pp. 94–98; Allen French, "The Arms and Military Training of Our Colonizing Ancestors," MHS, LXVII (Boston, 1945), pp. 18–19; Theodore Ropp, *War in the Modern World* (New York, 1962), pp. 50–51; and Harold L. Peterson, *Arms and Armor in Colonial*

America 1526–1783 (Harrisburg, PA., 1956), p. 160.

16. See Richard Ward, *Animadversions of Warre* (London, 1639), p. 261; Firth, *Cromwell's Army*, pp. 88–89 fn.; Boynton, *Elizabethan Militia*, pp. 113–114; Webb, *Elizabethan Military Science*, pp. 93–98; A.R. Hall, *Ballistics in the Seventeenth Century* (Cambridge, England, 1952), p. 8; B.H. Liddell Hart, "Armed Forces and the Art of War: Armies" in *The Zenith of European Power 1830–1870*, *The New Cambridge Modern History*, X (Cambridge, Eng., 1967), p. 303; and Stephen V. Granesay, "The Craft of the Early American Gunsmith," *Bulletin of the Metropolitan Museum of Art*, VI, no. 2 (October, 1947), p. 61.

17. See Peterson, *Arms and Armor*, p. 14; Lord Orrery, *A Treatise of the Art of War* (1677) quoted in Grose, *Military Antiquities*, I, p. 153 fn.; Grose, *Military Antiquities*, II, p. 336; Firth, *Cromwell's Army*, p. 90; John Scoffern, *Projectile Weapons of War and Explosive Compounds* (London, 1845), pp. 77, 77 fn.; John Metschl, "The Rudolph J. Nunnemacher Collection of Projectile Arms," *Bulletin of the Public Museum of the City of Milwaukee*, IX (1928), pp. 52–54; Carl P. Russell, *Guns on the Early Frontiers* (Berkeley, 1957), pp. 5, 6, 10; Harold L. Peterson, "The Military Equipment of the Plymouth and Bay Colonies 1620–1690," *New England Quarterly*, XX, no. 2 (1947), p. 202; and Charles Winthrop Sawyer, *Firearms in American History 1600–1800* (Boston, 1910), p. 6. Sawyer and Peterson say that musketeers tilted their weapons to aid ignition of the charge and therefore could not aim well. This appears to be a misreading of Johann Jacob Von Wallhausen, *Kriegskunst Zu Fuss*, I (Frankfurt–on–Main, 1615), which is quoted in Russell, *Guns on the Early Frontiers*, p. 5, and in Metschl, "Nunnemacher Collection," p. 52. Wallhausen says one should tilt and tap the gun after the priming powder is put in the pan in order to fill the touchhole. This is excellent advice, but it is done *before* you aim and fire. Tilting a musket in firing would be not only unnecessary, inaccurate, and awkward, but also painful when the weapon recoiled.

18. See French, "Arms and Military Training," pp. 3–10, 14–21; Barnes, *Somerset*, pp. 112–113; Firth, *Cromwell's Army*, pp. 69–71, 118–119; and Grose, *Military Antiquities*, I, p. 348.

19. See French, "Arms and Military Training," pp. 3, 8, 14–15, 21; Firth, *Cromwell's Army*, pp. 9–10; *Instructions for Musters and Armes*, pp. 6, 11–12; Bariffe, *Military Discipline*, pp. 3–5; and Elton, *Art Military*, pp. 5–6.

20. See *Watertown Records* (3 vols.; Watertown, MA., 1894–1904), I, p. 38; Douglas Edward Leach, *Flintlock and Tomahawk: New England in King Philip's War* (New York, 1966), pp. 11–12; NHCR, I, p. 202, II, p. 174; and Roger Williams, *A Key into the Language of America* (Providence, R.I., 1936), pp. 90–91 [1643].

21. See Benjamin Church, *The History of King Philip's War*, ed. by Henry Martyn Dexter (Boston, 1865), p. 54; Elton, *Art Military*, pp. 5–6; Bariffe, *Military Discipline*, pp. 3–5; Sylvester Judd, *History of Hadley* (Springfield, MA., 1905), p. 215; John Henry Stanley, "Preliminary Investigation of Military Manuals of American Imprint Prior to 1800" (M.A. thesis, Brown University, 1964), p. 2; and George Madison Bodge, *Soldiers in King Philip's War* (Leominster, MA., 1896), pp. 477–479.

22. John Underhill, "Newes from America," *MHC*, VI, 3rd series (Boston, 1837), p. 23.

23. *Ibid.*, p. 27.

24. See William Bradford, *Of Plymouth Plantation 1620–1647*, ed. by S.E. Morison (New York, 1967), p. 70; and William Brigham, ed., *The Compact With the Charter and Laws of the Colony of New Plymouth* (Boston, 1836), p. 184.

25. *CCR*, II, pp. 451–452.

26. See the Reverend William Hubbard, *The History of the Indian Wars in New England*, ed. by Samuel G. Drake (2 vols.; Roxbury, MA., 1865), I, p. 113; and MA, LXVII, p. 263.

27. See Bergstrom, "English Game Laws," pp. 689–690; and Alden Vaughan, *New England Frontier: Puritans and Indians 1620–1675* (Boston, 1965), p. 108.

28. See John Pory, "John Pory to the Earl of Southampton," in *Three Visitors to Early Plymouth*, ed. by Sydney V. James, Jr. (Plymouth, MA., 1963), p. 10; and Thomas Morton, *New English Canaan* (New York, 1967), pp. 202–203, 232.

29. See William Wood, *New England's Prospect* (Boston, 1865), pp. 35, 100 [1634]; and Williams, *Key*, p. 90.

30. See John Pory, "To the Governor of Virginia," in James, ed., *Three Visitors*, p. 16; Williams, *Key*, pp. 90–91; Wood, *New England's Prospect*, pp. 97–99; Edward Winslow, *Hypocrisie Unmasked* (Providence, 1916), p. 82; William Bradford, "A Descriptive and Historical Account of New England in Verse," *MHC*, III, 1st series (Boston, 1810), p. 82; *PCR*, IX, p. 209; Adrien Van Der Donck et al., "The Representation of New Netherland," in *Narratives of New Netherland 1609–1664*, ed. by J. Franklin Jameson (New York, 1959), p. 303; and John Josselyn, "An Account of Two Voyages to New England," *MHC*, III, 3rd series (Cambridge, MA., 1833), p. 309.

31. See Josselyn, "An Account of Two Voyages," p. 270; Bradford, "A Descriptive and Historical Account of New England in Verse," p. 82; Wood, *New England's Prospect*, p. 99; *CCR*, I, p. 74; Morton, *New English Canaan*, pp. 202–203; Vaughan, *New England Frontier*, p. 108; and *RICR*, I, p. 62.

32. See Bradford, *Of Plymouth Plantation*, p. 207; *MCR*, I, pp. 118, 127; and Brigham, ed., *Laws of New Plymouth*, p. 94.

33. See Wood, *New England's Prospect*, pp. 26–27, 218; *PCR*, III, p. 50, IV, p. 6; *MCR*, IV–2, p. 2; *CCR*, I, p. 561; Brigham, ed., *Laws of New Plymouth*, pp. 93–94, 98; *RICR*, I, pp. 124–125; and *NHCR*, I, pp. 182, 217–218.

34. See Isaack de Rasieres, "Letter of Isaack de Rasieres to Samuel Blommaert, 1628(?)," in *Narratives of New Netherland 1609–1664*, ed. by J. Franklin Jameson (New York, 1959), p. 115; Wood, *New England's Prospect*, p. 34; Morton, *New English Canaan*, p. 190; and *Mourt's Relation*, p. 26.

35. See Wood, *New England's Prospect*, pp. 34–35; and de Rasieres, "Letter," p. 114.

36. See Wood, *New England's Prospect*, pp. 31–32; and Josselyn, "An Account of Two Voyages," p. 278.

37. See Wood, *New England's Prospect*, p. 24; and Josselyn, "An Account of Two Voyages," p. 272.

38. Wood, *New England's Prospect*, pp. 52–53, 57.

39. See Wood, *New England's Prospect*, pp. 52–53; Edward Johnson, *Johnson's Wonder-Working Providence*, ed. by J. Franklin Jameson (New York, 1910), p. 83; and Bradford, *Of Plymouth Plantation*, pp. 101–102, 109–110.

40. Edward Winslow, "Good Newes from New England" in *Chronicles of the Pilgrim Fathers of the Colony of Plymouth 1602–1625*, ed. by Alexander Young (Boston, 1841), p. 294.

41. See Bradford, *Of Plymouth Plantation*, p. 90; Pond letter in *Winthrop Papers*, III, p. 18; *Winthrop Papers*, IV, p. 492; and *MCR*, I, p. 109.

42. See Wood, *New England's Prospect*, p. 80; and *Winthrop Papers*, IV, p. 492.

43. James Deetz, "The Reality of the Pilgrim Fathers," *Natural History*, LXXVIII, No. 9 (November, 1969), p. 43. Also conversations with the author.

44. Williams, *Key*, pp. 90–91.

45. From a legend collected by Frank Speck and quoted in William S. Simmons, *Spirit of the New England Tribes: Indian History and Folklore, 1620–1684* (Hanover, NH, 1986), p. 96. The author is grateful to Professor Simmons for demonstrating the survival of seventeenth-century Indian legends and for pointing out how folklore could inform this particular analysis of Indian military practices.

46. See Russell, *Guns on the Early Frontiers*, pp. 10–11, 14; Trelease, *Indian Affairs in Colonial New York* (Ithaca, N.Y., 1960), p. 146; Judd,

History of Hadley, p. 183; Increase Mather, *Early History of New England*, ed. by Samuel G. Drake (Boston, 1864), p. 206; and CA, War Council I, p. 44.

47. See Samuel Sewall, "Sewall Papers 1674–1700," *MHC*, VI, 5th series (Boston, 1879), p. 45; and Bodge, *Soldiers in King Philip's War*, pp. 45–46. Bodge, who is not reliable on weapons, says the Indians "used slugs or heavy shot instead of bullets." The Indians did use bullets as well as shot. Molds for shot and for heavy musket balls have been found at a number of Indian sites in southern New England. For analysis of the shot molds found at the Fort Shantok site, see Lorraine Williams, "Ft. Shantok and Ft. Corchaug: A Comparative Study of Seventeenth Century Culture Contact in the Long Island Sound Area" (Ph.D. dissertation, New York University, 1972), pp. 333–335.

48. See Winslow, "A Letter," p. 86; and John Winthrop, *Winthrop's Journal*, ed. by James Kendall Hosmer (2 vols.; New York, 1908), I, p. 83.

49. Hubbard, *History of the Indian Wars*, II, p. 36.

50. See *NHCR*, I, pp. 26, 96, 131; *CCR*, I, p. 544; and *MCR*, IV, p. 44.

NOTES TO CHAPTER V

1. William Wood, *New England's Prospect* (Boston, 1865), pp. 87–88.

2. See Ebenezer Hazard, ed., *Historical Collections* (Philadelphia, 1792, 1794), I, p. 312; and William Bradford, *Of Plymouth Plantation 1620–1647*, ed. by S.E. Morison (New York, 1967), p. 207.

3. See Thomas Shepard, "The Clear Sun-Shine of the Gospel Breaking Forth Upon the Indians in New-England," *MHC*, IV, 3rd series (Cambridge, 1834), p. 65; Wood, *New England's Prospect*, pp. 87–88; and "Acculturation: An Exploratory Formulation," Social Science Research Council Summer Seminar on Acculturation, 1953, in *American Anthropologist*, 56 (1954), p. 990.

4. See Daniel Gookin, "Historical Collections of the Indians in New England," *MHC*, I, 1st series (Boston, 1792), pp. 207, 210; Wood, *New England's Prospect*, p. 69; Roger Williams, *A Key into the Language of America* (Providence, R.I., 1936), pp. 38–39, 44–45; and MA, LXVIII, p. 68a. Patricia Rubertone of the Brown University Anthropology Department and Paul Robinson of the Rhode Island Historical Preservation Commission have speculated that the African-American population in Rhode Island may have been one of the various sources of technological information for the Narragansetts after the 1650s. In many areas of Africa, metallurgical capabilities were already advanced by this period.

Both African slaves and freed slaves did work as craftsmen in some Rhode Island settlements before King Philip's War and may have had considerable contact with Native Americans.

5. Gookin, "Historical Collections," pp. 162, 166–167.

6. See *PCR*, X, p. 251; and John Eliot, "The Day Breaking if Not the Sun Rising of the Gospel With the Indians in New England," *MHC*, IV, 3rd series (Cambridge, 1834), p. 18.

7. See Gookin, "Historical Collections," pp. 207, 210; John Eliot and Thomas Mayhew, "Tears of Repentance; Or, A Further Narrative of the Progress of the Gospel Amongst the Indians in New England," *MHC*, IV, 3rd series (Cambridge, 1834), p. 50; John Eliot, 1647 letter in Shepard, "Clear Sun-Shine of the Gospel," p. 50; and Ola Winslow, *John Eliot* (Boston, 1968), p. 161.

8. See Gookin, "Historical Collections," pp. 212–213; *PCR*, X, pp. 106, 219; Winslow, *John Eliot*, pp. 116–118; and Alden Vaughan, *New England Frontier: Puritans and Indians 1620–1675* (Boston, 1965), pp. 161, 303, 308.

9. See John Eliot, 1650 letter in Henry Whitfield, "The Light Appearing More and More Towards the Perfect Day; Or, A Further Discovery of the Present State of the Indians in New England," *MHC*, IV, 3rd series (Cambridge, 1834), pp. 138,142–143; 1651 letters by John Eliot, John Wilson and John Endecott in Henry Whitfield, "Strength Out of Weaknesse; Or, A Glorious Manifestation of the Further Progresse of the Gospel Among the Indians of New England," *MHC*, IV, 3rd series (Cambridge, 1834), pp. 167, 174, 177, 190–191; Gookin, "Historical Collections," pp. 184, 206–207, 210, 213; and Daniel Gookin, "An Historical Account of the Doings and Sufferings of the Christian Indians in New England in the Years 1675, 1676, and 1677," *Transactions and Collections of the American Antiquarian Society*, II (Cambridge, 1836), p. 512.

10. See John Eliot, "John Eliot to Robert Boyle, October 23, 1677," *MHC*, III, 1st series (Boston, 1794), p. 179; Gookin, "Christian Indians in New England," p. 462; William Harris, *A Rhode Islander Reports on King Philip's War: The Second William Harris Letter of August 1676*, ed. by Douglas Edward Leach (Providence, R.I., 1963), pp. 64, 66; and MA, LXVIII, p. 136.

11. See John Underhill, "Newes From America," *MHC*, VI, 3rd series (Boston, 1837), p. 17; and Bradford, *Of Plymouth Plantation*, p. 207.

12. See Bradford, *Of Plymouth Plantation*, p. 207; and William Bradford, "A Descriptive and Historical Account of New England in Verse," *MHC*, III, 1st series (Boston, 1810), p. 82.

13. See Bradford, *Of Plymouth Plantation*, p. 207; Williams, *Key*, pp. 44–45; Nathaniel Saltonstall, "A New and Further Narrative," in

Narratives of the Indian Wars, ed. by Charles H. Lincoln (New York, 1959), p. 96; Lorraine Williams, "Ft. Shantok and Ft. Corchaug: A Comparative Study of Seventeenth Century Culture Contact in the Long Island Sound Area" (Ph.D. dissertation, New York University, 1972), pp. 334–335; Jean-Francois Blanchette, "Firearms," in Susan Gibson, ed., *Burr's Hill: A 17th Century Wampanoag Burial Ground in Warren, Rhode Island* (Providence, RI, 1980), p. 70; and Charles C. Willoughby, *Antiquities of the New England Indians* (Cambridge, 1935), pp. 183, 235, 243. See also a stone mold in the collection of the Robbins Museum, Middleboro, MA.

14. See Williams, "Ft. Shantok and Ft. Corchaug," pp. 333–335; James W. Bradley, *Evolution of the Onondaga Iroquois: Accommodating Change, 1500–1655* (Syracuse, 1987), pp. 175–176; and Blanchette, "Firearms," pp. 68–70.

15. See Bradford, "New England in Verse," p. 82; MCR, II, p. 163; PCR, IX, pp. 21–22; Harris, *A Rhode Islander Reports on King Philip's War*, p. 64; Nathaniel Saltonstall, "A Continuation of the State of New England," in *Narratives of the Indian Wars*, p. 59; Saltonstall, "A New and Further Narrative," p. 96; Joseph R. Mayer, *Flintlocks of the Iroquois 1620–1687*, Research Records of Rochester Museum of Arts and Sciences, no. 6 (Rochester, N.Y., 1943), pp. 33–34; T.M. Hamilton, "Some Gun Parts from 17th Century Seneca Sites," in *Indian Trade Guns*, ed. T.M. Hamilton, *The Missouri Archaeologist*, 22 (Columbia, MO., 1960), pp. 103–107; Jean-Francois Blanchette, "Firearms," in Susan G. Gibson, *Burr's Hill* (Providence, R.I., 1980), pp. 68–70; and Kenneth L. Feder, "Of Stone and Metal: Trade and Warfare in Southern New England," *The New England Social Studies Bulletin*, vol. 44, no. 1 (1986), p. 33.

16. See PCR, X, p. 17; and NHCR, II, p. 594.

17. MA, XXX, pp. 63–64.

18. See "Testimony of Hugh Cole," at Plymouth Mar. 8, 1670/1, Misc. bound MSS, Mass. Historical Society; and Saltonstall, "A Continuation," p. 59.

19. See Paul Robinson, Marc A. Kelley, and Patricia Rubertone, "Preliminary Biocultural Interpretations from a Seventeenth-Century Narragansett Indian Cemetery in Rhode Island," in *Cultures in Contact: The Impact of European Contacts on Native American Cultural Institutions, A.D. 1000–1800*, ed. by William W. Fitzhugh, (Washington, 1985), pp. 111, 120; and Hamilton, "Some Gun Parts From 17th Century Seneca Sites," pp. 101–107.

20. See Edward Johnson, *Johnson's Wonder-Working Providence*, ed. by J. Franklin Jameson (New York, 1910), pp. 149–150; John Mason, "A Brief History of the Pequot War," MHC, VIII, 2nd series

(1819), p. 134. On later powder shortages, see, for example, Richard LeBaron Bowen, *Early Rehoboth* (4 vols.; Concord, N.H., 1945–50), III, 96.

21. See Bradford, *Of Plymouth Plantation*, p. 207; Harris, *A Rhode Islander Reports on King Philip's War*, pp. 64, 66; Arthur Pine Van Gelder and Hugo Schlatter, *The History of the Explosives Industry in America* (New York, 1927), pp. 8–14, 29–35, 37–39; MCR, II, pp. 17–18, 29, 102, IV–2, pp. 296–297, 320, V, pp. 51, 64, 72.

22. See Wood, *New England's Prospect*, p. 94; Willoughby, *Antiquities of the New England Indians*, pp. 284–288; Peter Thomas, "In the Maelstrom of Change: The Indian Trade and Cultural Process in the Middle Connecticut River Valley, 1635–1665" (Ph.D. dissertation, University of Massachusetts, 1979), pp. 77, 115–117; and Samuel de Champlain, *The Works of Samuel de Champlain*, ed. by H.P. Biggar (6 vols.; Toronto, 1922–36), I, pp. 329–330.

23. See MCR, II, 72; John W. DeForest, *History of the Indians of Connecticut* (Hamden, CT., 1964), pp. 213–214; H.M.C., "Pomham and His Fort," *RIHC*, XI (1918), p. 33; Williams, "Ft. Shantok and Ft. Corchaug," pp. 15, 77, 159, 196; and Gookin, "Christian Indians in New England," p. 436. See Robert Ward, *Animadversions of Warre* (London, 1639) for coverage of fortifications and military engineering in general.

24. See the Rev. William Hubbard, *The History of the Indian Wars in New England*, ed. by Samuel G. Drake (Roxbury, MA., 1865), I, pp. 146–147; Saltonstall, "A Continuation," p. 58; and Douglas Edward Leach, *Flintlock and Tomahawk: New England in King Philip's War* (New York, 1966), p. 19. Leach's study is the standard work on King Philip's War and was a great help in the research for this book.

25. Saltonstall, "A Continuation," p. 59.

26. Saltonstall, "A New and Further Narrative," p. 96.

27. See Sidney S. Rider, *The Lands of Rhode Island As They Were Known to Caunounicus and Miantunnomu* (Providence, 1904), pp. 236–244; and Elisha R. Potter, "The Early History of Narragansett," *RIHC*, III (Providence, 1835), 84 note. The Rhode Island Historical Preservation Commission has a detailed archaeological map of the fort site prepared for the Commission by M. Stachiw, S. Cole, and G. Gustafson.

28. See Saltonstall, "A New and Further Narrative," p. 96; and Rider, *The Lands of Rhode Island*, pp. 236–238, 241–242.

29. Saltonstall, "A New and Further Narrative," p. 96.

30. See Stuart D. Goulding, "Deep in the Rhode Island Forest," *Yankee*, XXXIII (Mar. 1969), pp. 44–46 for some of the legends surrounding Stonewall John. C.H. Lincoln, editor of Nathaniel

Saltonstall's wartime accounts, ignores the legend of the renegade military engineer but suggests that a colonial turncoat, Joshua Tift, may have helped John design the fort in the Great Swamp. See Lincoln, ed., *Narratives*, pp. 48 fn, 49 fn. Actually, Tift (or Tefft in some sources) had only been with the Narragansetts a short while. Although he admitted to helping in the construction of the fort, he was not a skilled craftsman nor an expert on military fortifications. Tift, who had once served briefly in a militia unit, probably became part of the Narragansett work force led by Stonewall John. He was later executed as a traitor by colonial authorities. For information on Tift, see Colin Calloway, "Rhode Island Renegade: The Enigma of Joshua Tefft," *Rhode Island History*, XLIV, no. 4 (Nov., 1984), pp. 137–145.

31. See Richard Slotkin, *Regeneration Through Violence* (Middletown, CT., 1973), pp. 69–88; John Ferling, "The New England Soldier: A Study in Changing Perceptions," in *American Quarterly*, XXX, no. 1 (Spring, 1981), pp. 30–33; James Axtell, "The Scholastic Philosophy of the Wilderness," in *William and Mary Quarterly*, XXIX, no. 3, 3rd series (July, 1972), pp 343–344; and Underhill, "Newes from America," p. 15. For examples of excesses in continental warfare, see C.V. Wedgwood, *The Thirty Years War* (N.Y., 1961). Francis Jennings, *The Invasion of America: Indians, Colonialism, and the Cant of Conquest* (Chapel Hill, N.C., 1975), pp. 5–7, 212–213, discusses religious and racial justifications for massacres and compares English brutality in Ireland in the sixteenth and seventeenth centuries with treatment of Indian opponents. Adam J. Hirsch, "The Collision of Military Cultures in Seventeenth-Century New England," *Journal of American History*, LXXIV, no. 4 (March, 1988), pp. 1197–1209, casts new light on the issue of "military acculturation." He argues that the experience of warfare against Indians whose military culture was so different from that of the Europeans promoted the loosening of restraints and the hardening of attitudes that produced the massacre at the Pequot fort: "It was largely the Indians' disapproval and avoidance of the debilitating modes of Old World warfare that led the settlers to employ new strategies and tactics even more murderous than the original ones."

32. See John Mason, "A Brief History of the Pequot War," pp. 133–138; John Underhill, "Newes from America," pp. 15, 23; and Phillip Vincent, "A True Relation of the Late Battle Fought in New England," *MHC*, VI, 3rd series (Boston, 1837), pp. 36–40.

33. See Mason, "Brief History of the Pequot War," pp. 138–141; Underhill, "Newes from America," pp. 23–25; and Vincent, "A True Relation of the Late Battle," pp. 36–40. For the most vigorous modern condemnation of the English actions at the Pequot fort, see Jennings, *The Invasion of America*, pp. 220–225.

34. See Roger Williams to Gov. Vane and Lt. Gov. Winthrop, May 15, 1637, in *Winthrop Papers 1498–1649* (5 vols.; Boston, 1929–1947), III, p. 414; and Underhill, "Newes from America," p. 27.

35. Underhill, "Newes from America," p.25.

36. *Ibid.*

37. Vincent, "A True Relation of the Late Battle," p. 42.

38. Thomas Wheeler, "Captain Thomas Wheeler's Narrative, 1675," *Old South Leaflets*, no. 155 (Boston), pp. 4–12. For other descriptions of attempts to burn houses see Gookin, "Christian Indians in New England," p. 493; Saltonstall, "A New and Further Narrative," p. 79; MA, LXVIII, p. 49; and Benjamin Tompson, *New England's Crisis* (Boston, 1676), p. 24. Leach, *Flintlock and Tomahawk* contains much information on the ferocity of the war.

39. Wheeler, "Wheeler's Narrative," pp. 1–4.

40. See Increase Mather, *A Brief History of the War With the Indians in New England*, ed. by Samuel G. Drake (Albany, N.Y., 1862), pp. 84–85; and Hubbard, *History of the Indian Wars*, I, pp. 112–115.

41. See Hubbard, *History of the Indian Wars*, I, p. 210; and Mather, *Brief History of the War With the Indians*, p. 136.

42. See Gookin, "Christian Indians in New England," pp. 441–442; and Benjamin Church, *The History of King Philip's War* [dictated to his son, Thomas Church, probably in 1715, and published in 1716], ed. by Henry Martyn Dexter (Boston, 1865), p. 132.

43. See Church, *History of King Philip's War*, pp. 67–69, 67 fn.; and George D. Langdon, Jr., *Pilgrim Colony: A History of New Plymouth 1620–1691* (New Haven, 1966), p. 180.

44. See CA, War Colonial I, p. 66; CCR, II, pp. 424, 438; and Gookin, "Christian Indians in New England," p. 437.

45. See Harris, *A Rhode Islander Reports on King Philip's War*, pp. 64, 66; Gookin, "Christian Indians in New England," p. 462; MA, XXX, pp. 193–193a, LXVII, p. 213; and "Winthrop Papers," *MHC*, I, 5th series, p. 106.

46. See Gookin, "Christian Indians in New England," pp. 441–450; MCR, V, p. 57; MA, XXX, pp. 186, 187, 190, 193, 193a, 195, 198b; and "Letters from the Reverend John Eliot to Honorable Robert Boyle," *MHC*, III, 1st series (Boston, 1794), p. 179.

47. "Winthrop Papers," *MHC*, I, 5th series, p. 106. For a similar rumor see Harris, *A Rhode Islander Reports on King Philip's War*, pp. 64–66.

48. Gookin, "Christian Indians in New England," pp. 477, 480, 500–502.

49. See MA, LXVII, p. 275; and Hubbard, *History of the Indian Wars*, I, pp. 106, 178.

50. See Mather, *Brief History of the War With the Indians*, p. 121; and Gookin, "Christian Indians in New England," pp. 448, 505.

51. See Gookin, "Christian Indians in New England," pp. 506–513, 524–525; *MCR*, V, pp. 84, 85, 87; CA, War Colonial I, p. 78; and MA, XXX, pp. 201, 204, 212b, LXVIII, pp. 220–228.

52. See Mather, *Brief History of the War With the Indians*, p. 143; and Gookin, "Christian Indians in New England," pp. 513, 517–519.

53. See Bowen, *Early Rehoboth*, III, pp. 14–15; Mather, *Brief History of the War With the Indians*, pp. 127–128; Langdon, *Pilgrim Colony*, p. 180; Church, *History of King Philip's War*, pp. 67–69; and MA, LXVIII, p. 220.

54. See Gookin, "Christian Indians in New England," p. 441; Benjamin Tompson, *New England's Crisis* (Boston, 1676), pp. 12–13; John and William Pynchon, "The Pynchon Court Records," in *Colonial Justice in Western Massachusetts (1639–1702)*, ed. by Joseph H. Smith (Cambridge, MA., 1961), p. 39; and Harris, *A Rhode Islander Reports on King Philip's War*, p. 28.

55. See Richard Elton, *The Compleat Body of the Art Military* (London, 1659), p. 184; Charles H. Lincoln, ed., *Narratives of the Indian Wars* (New York, 1959), pp. 39, 58–59; Leach, *Flintlock and Tomahawk*, p. 105; and George Madison Bodge, *Soldiers in King Philip's War* (Leominster, MA., 1896), pp. 190–191.

56. Hubbard, *History of the Indian Wars*, I, pp. 113–115.

57. MA, LXVII, p. 208.

58. See Church, *History of King Philip's War*, pp. 33–36; "Lt. Thomas to Gov. Winslow, August 10, 1675," in Bowen, *Early Rehoboth*, III, pp. 95–96; and John Russell quoted in Mather, *Brief History of the War With the Indians*, p. 77.

59. MA, LXII, p. 263.

60. Church, *History of King Philip's War*, pp. 100, 103–104, 116, 121–123.

61. *Ibid.*, pp. 103–104, 112, 120–123, 143, 161–62. See also the analysis in Axtell, "The Scholastic Philosophy of the Wilderness," pp. 346–350.

62. Church, *History of King Philip's War*, p. 123.

63. *Ibid.*, pp. 143–148.

64. See Leach, *Flintlock and Tomahawk*, pp. 219–220, 242–243 for his conclusions on why the Indians lost the war. The defensive strength of the fortified garrison houses in New England towns should also be considered a major factor.

65. See Harris, *A Rhode Islander Reports on King Philip's War*, pp. 60–63; Winthrop MSS, Massachusetts Historical Society, XVII, p. 13, XVIII, p. 138; "Winthrop Papers," pp. 106, 109; Lincoln, ed.,

Narratives of the Indian Wars, pp. 73, 97–98, 160; MA, LXVII, pp. 256a, 275, LXVIII, pp. 49, 211, 241; *CCR*, II, pp. 450, 467; Samuel Sewall, "Sewall Papers 1674–1700," *MHC*, V, 5th series (Boston, 1878), p. 14; Church, *History of King Philip's War*, pp. 111–112, 117–118, 132, 163, 170, 175–176; Mather, *Brief History of the War With the Indians*, pp. 140, 145; Gookin, "Christian Indians in New England," pp. 457, 504–505; Wilcomb Washburn, "Seventeenth-Century Indian Wars," in Bruce Trigger, ed., *Northeast*, vol. 15 of William C. Sturtevant, ed., *Handbook of North American Indians* (Washington, D.C., 1978–1984), pp. 94, 98–100; Neal Salisbury, "Toward the Covenant Chain: Iroquois and Southern New England Algonquians, 1637–1684," in Daniel Richter and James H. Merrell, eds., *Beyond the Covenant Chain: The Iroquois and Their Neighbors in Indian North America, 1600–1800* (Syracuse, 1987), pp. 70–72; and James Axtell, *The European and the Indian: Essays in the Ethnohistory of Colonial North America* (Oxford, 1981), pp. 146–148, 301–302.

66. See Axtell, *The European and the Indian*, pp. 150–151, 297–298; Irving Hallowell, "The Impact of the America Indian on American Culture," *American Anthropologist*, 59 (April, 1957), pp. 201–217; John L. Cotter, *Archaeological Excavations at Jamestown* (Washington, D.C., 1958), pp. 162–163; Wood, *New England's Prospect*, p. 76; and MA, LXVII, 135a.

67. John Eliot to Robert Boyle, October 23, 1677, *MHC*, III, 1st series (Boston, 1794), p. 178.

NOTES TO ILLUSTRATIONS

Fig. 1. Romantic view of the New England forest. Illustration from Francis S. Drake, *Indian History for Young Folks* (New York, 1885).

Fig. 2. Tribal map of southern New England. Map by Lyn Malone.

Fig. 3. Birchbark canoe. Illustration from Welcome Arnold Green, *The Providence Plantations for Two Hundred and Fifty Years* (Providence, 1886).

Fig. 4. Making a dug out canoe. An engraving of a 1585 painting by Roanoke colonist John White, published in Thomas Hariot, *Admiranda Narratio Fida Tamen . . .* , for Theodore de Bry (Frankfurt am Main, 1590). Courtesy of the John Carter Brown Library at Brown University.

Fig. 5. The fortified, coastal Algonquian town of Pomeiooc. An engraving of a 1585 painting by John White, published in Hariot, *Admiranda*. Courtesy of the John Carter Brown Library at Brown University.

Fig. 6. Palisaded Indian villages shown on several maps by John Seller, including his map of New England in 1676. This is the best example of his illustration, from the Seller Map of New Jersey, 1677. Courtesy of the John Carter Brown Library at Brown University.

Fig. 7. The Sudbury bow, 1660, in the collections of the Peabody Museum of Harvard University. Drawing by Lyn Malone, by permission of the Museum.

Fig. 8. An Indian hunting with a bow. Illustration from the Seller Map of New Jersey, 1677. Courtesy of the John Carter Brown Library at Brown University.

Fig. 9. Axe with smooth stone head set in original haft. Illustration by William S. Fowler, reprinted, by permission, from *Bulletin of the Massachusetts Archaeological Society,* XXI, no. 3–4 (1960).

Fig. 10. Grooved, stone axe head. Drawing of an artifact in the collections of the George Hail Free Library, Warren, Rhode Island. Courtesy of the Haffenreffer Museum of Anthropology at Brown University, by permission of the Library.

Fig. 11. Nineteenth-century engraving of an imagined Indian religious ceremony. Illustration from John W. Barber, *The History and Antiquities of New England, New York, New Jersey, and Pennsylvania* (Hartford, 1842).

Fig. 12. Warriors with bows. Illustration from Sarah S. Jacobs, *Nonantum and Natick* (Boston, 1853).

Fig. 13. Landing of the Pilgrims. Illustration from Barber, *The History and Antiquities of New England.*

Fig. 14. The "Great Mortality." Illustration from Barber, *The History and Antiquities of New England.*

Fig. 15. Trade hatchet in the collections of the Haffenreffer Museum. Drawing by Lyn Malone, by permission of the Museum.

Fig. 16. Axe head, from Burr's Hill. Illustration by Jean Blackburn, reprinted from Gibson, ed., *Burr's Hill* (Providence, R.I., 1980), by permission of the Haffenreffer Museum of Anthropology at Brown University and the George Hail Free Library.

Fig. 17. Knife blade with tang, from Burr's Hill. Illustration by Jean Blackburn, reprinted from Gibson, ed., *Burr's Hill,* by permission of the Haffenreffer Museum of Anthropology at Brown University.

Fig. 18. Operation of a matchlock musket. Drawing by P.D. Malone and the author. Based partly on a diagram in Jack O'Connor, *Complete Book of Rifles and Shotguns* (New York, 1961). Reprinted, by permission, from *American Quarterly,* XXV, no. 1 (March, 1973).

Fig. 19. Pilgrims in their first military action against Indians. Illustration from Drake, *Indian History for Young Folks.*

Fig. 20. Operation of a flintlock musket. Drawing by P.D. Malone and the author. Based partly on a diagram in O'Connor, *Complete Book of Rifles and Shotguns.* Reprinted, by permission, from *American Quarterly,* XXV, no. 1 (March, 1973).

Fig. 21. Dutch traders with Indians. Illustration from Drake, *Indian History for Young Folks.*

Fig. 22. Two strings of wampum, from Burr's Hill, in the collections of the George Hail Free Library. Drawing by Lyn Malone, by permission of the Library.

Fig. 23. Portrait of Ninigret II. Courtesy of the Museum of Art, Rhode Island School of Design, Gift of Mr. Robert Winthrop. For a scholarly discussion of this painting, and of other identified images of New England Indians, see William S. Simmons, "The Earliest Prints and Paintings of New England Indians," *Rhode Island History,* XLI, no. 3 (Aug., 1982), pp. 73–85.

Fig. 24. An Indian hunting deer with a firearm. Illustration from John Frost, *Pictorial History of Indian Wars and Captivities . . .* (New York, 1873).

Fig. 25. Flintlock from a seventeenth-century Wampanoag burial at Burr's Hill. Illustration by Robin Gibson, reprinted from Gibson, ed., *Burr's Hill,* by permission of the Haffenreffer Museum of Anthropology and the Museum of the American Indian. Richard Colton, an authority on firearms from this period, identifies this lock as almost certainly made in Holland or Belgium. For data on the identification and dating of Dutch trade muskets see Jan Piet Puype, *Dutch and other Flintlocks from Seventeenth Century Iroquois Sites,* Part I of *Proceedings of the 1984 Trade Gun Conference,* Research Records of the Rochester Museum and Science Center, no. 18 (Rochester, NY, 1984).

Fig. 26. Hunting deer in seventeenth-century England, using hounds. Illustration from George Turberville, *Turberville's Booke of Hunting, 1576* (Oxford, 1908).

Fig. 27. European musketeer. One of a series of illustrations showing the exercise of arms in Jacob De Gheyn, *Waffenhandlung Von Den Roren, Musquetten, Undt Spiessen* (Amsterdam, 1608). Illustration courtesy of the Anne S.K. Brown Military Collection, Brown University.

Fig. 28. One of the recommended steps in loading a musket. Illustration from De Gheyn, *Waffenhandlung.* Courtesy of the Anne S.K. Brown Military Collection, Brown University.

Fig. 29. Blowing on the tip of the match. Illustration from De Gheyn, *Waffenhandlung.* Courtesy of the Anne S.K. Brown Military Collection, Brown University.

Fig. 30. Preparing to "present" the musket in the direction of the enemy. Illustration from De Gheyn, *Waffenhandlung.* Courtesy of the Anne S.K. Brown Military Collection, Brown University.

Fig. 31. Presenting (leveling) the musket and firing on command. Illustration from De Gheyn, *Waffenhandlung.* Courtesy of the Anne S.K. Brown Military Collection, Brown University.

Fig. 32. Pikemen and musketeers shown in a seventeenth-century military manual. Illustration from Johan Boxel, *Exercitie Memorie Van De Compagnie Guardes* (The Hague, Netherlands, 1669). Courtesy of the Anne S.K. Brown Military Collection.

Fig. 33. Musketeers, with pikemen in position behind them, fire a volley. Drawing by John Lang from a photograph by Ted Avery.

Fig. 34. An Indian and a musket. Illustration from Jacobs, *Nonantum and Natick*.

Fig. 35. John Eliot preaching to the Indians. Illustration from Drake, *Indian History for Young Folks*.

Fig. 36. Indians building a meetinghouse at Natick. Illustration from Jacobs, *Nonantum and Natick*.

Fig. 37. Half of an Indian mold for casting six lead shot. Drawing by Lyn Malone, by permission of the Robbins Museum of the Massachusetts Archaeological Society.

Fig. 38. Half of a shot mold found at Burr's Hill. Illustration by Robin Gibson, reprinted from Gibson, ed., *Burr's Hill*, by permission of the Haffenreffer Museum of Anthropology at Brown University and the Museum of the American Indian.

Fig. 39. Both halves of a hinged shot mold found at Burr's Hill. Illustration by Robin Gibson, reprinted from Gibson, ed., *Burr's Hill*, by permission of the Haffenreffer Museum of Anthropology at Brown University and the Museum of the American Indian.

Fig. 40. Two common types of European gunflints. Illustration reprinted from Gibson, ed., *Burr's Hill*, by permission of the Haffenreffer Museum of Anthropology at Brown University.

Fig. 41. Indians repairing firearms. Illustration from Jacobs, *Nonantum and Natick*.

Fig. 42. A claw hammer from Burr's Hill. Illustration by Robin Gibson, reprinted from Gibson, ed., *Burr's Hill*, by permission of the Haffenreffer Museum of Anthropology at Brown University and the Museum of the American Indian.

Fig. 43. The Great Swamp Fight in 1675. Illustration from Drake, *Indian History for Young Folks*.

Fig. 44. Archaeological map of the Queen's Fort. Drawn by Myron Stachiw, S. Cole, and Gail Gustafson. Courtesy of the Rhode Island Historical Preservation Commission.

Fig. 45. A contemporary diagram of the attack on the Pequots' fort near the Mystic River in 1637. Illustration from John Underhill, *News from America* (London, 1638). Courtesy of the New York Public Library.

Fig. 46. The massacre at the Pequot fort. Illustration from Frost, *Pictorial History of Indian Wars*.

Fig. 47. Lancaster under attack. Illustration from Drake, *Indian History for Young Folks.*

Fig. 48. The Indian assault on Brookfield. Illustration from Frost, *Pictorial History of Indian Wars.*

Fig. 49. Use of a siege device at Brookfield. Illustration from Barber, *The History and Antiquities of New England.*

Fig. 50. An ambush at "Bloody Brook," near Deerfield. Illustration from Barber, *The History and Antiquities of New England.*

Fig. 51. A scout. Illustration from Frost, *Pictorial History of Indian Wars.*

Fig. 52. A Connecticut unit attacking Indians in a swamp. Illustration from Frost, *Pictorial History of Indian Wars.*

Fig. 53. A raid on a family in the fields. Illustration from Greene, *The Providence Plantations.*

Fig. 54. The attack on the Narragansetts' fort in the Great Swamp. Illustration from Barber, *The History and Antiquities of New England.*

Fig. 55. Portrait of Governor John Leverett wearing a "buff suit." Courtesy of the Essex Institute in Salem, MA. A buff coat of thick leather, identified as Leverett's and apparently the same as the one in the portrait at the Essex Institute, is in the collections of the Massachusetts Historical Society.

Fig. 56. An ambush in the forest. Illustration from Drake, *Indian History for Young Folks.*

Fig. 57. Captain Benjamin Church and his men adopting Indian tactics. Illustration from Drake, *Indian History for Young Folks.*

Fig. 58. Captain Church, with an Indian scout. Illustration from Daniel Strock Jr., *Pictorial History of King Philip's War* (Boston, 1852). Courtesy of the John Hay Library at Brown University.

Fig. 59. Paul Revere's fanciful engraving of the great Wampanoag sachem Metacomet. Illustration from Benjamin Church, *The Entertaining History of King Philip's War* (Newport, 1772). Courtesy of the John Carter Brown Library at Brown University.

Fig. 60. An Indian firing the shot that killed Metacomet. Illustration from S.G. Goodrich, *A Pictorial History of America* (Hartford, 1851).

Fig. 61. Defense of a garrison house. Illustration from Drake, *Indian History for Young Folks.*

Fig. 62. The capture of the Wampanoag war leader Annawon. Illustration from Barber, *The History and Antiquities of New England.*

Fig. 63. Indians in hiding. Illustration from Frost, *Pictorial History of Indian Wars.*

Index